Ben Stacy Jerrik (Ed.)

Roman Maev

Ben Stacy Jerrik (Ed.)

Roman Maev

Nikolay Basov

Part Press

Imprint

Permission is granted to copy, distribute and/or modify this document under the terms of the GNU Free Documentation License, Version 1.2 or any later version published by the Free Software Foundation; with no Invariant Sections, with the Front-Cover Texts, and with the Back- Cover Texts. A copy of the license is included in the section entitled "GNU Free Documentation License".

All parts of this book are extracted from Wikipedia, the free encyclopedia (www.wikipedia.org).

You can get detailed informations about the authors of this collection of articles at the end of this book. The editors (Ed.) of this book are no authors. They have not modified or extended the original texts.

Pictures published in this book can be under different licences than the GNU Free Documentation License. You can get detailed informations about the authors and licences of pictures at the end of this book.

The content of this book was generated collaboratively by volunteers. Please be advised that nothing found here has necessarily been reviewed by people with the expertise required to provide you with complete, accurate or reliable information. Some information in this book maybe misleading or wrong. The Publisher does not guarantee the validity of the information found here. If you need specific advice (f.e. in fields of medical, legal, financial, or risk management questions) please contact a professional who is licensed or knowledgeable in that area.

Any brand names and product names mentioned in this book are subject to trademark, brand or patent protection and are trademarks or registered trademarks of their respective holders. The use of brand names, product names, common names, trade names, product descriptions etc. even without a particular marking in this works is in no way to be construed to mean that such names may be regarded as unrestricted in respect of trademark and brand protection legislation and could thus be used by anyone.

Cover image: www.ingimage.com
Concerning the licence of the cover image please contact ingimage.

Publisher:
Part Press is a trademark of
International Book Market Service Ltd., 17 Rue Meldrum, Beau Bassin, 1713-01 Mauritius
Email: info@bookmarketservice.com
Website: www.bookmarketservice.com

Published in 2012

Printed in: U.S.A., U.K., Germany. This book was not produced in Mauritius.

ISBN: 978-613-8-98814-4

Contents

Articles

References

Roman_Maev

Roman Grigorievich Maev (Russian: Роман Григорьевич Маев), Ph.D., D.Sc., Prof. (born 1945 in Moscow) is a Russian-born physicist and the Founding Director of the Institute for Diagnostic Imaging Research in Windsor, Ontario, Canada.[1] He is the University of Windsor's Distinguished Professor of the Department of Physics and one of Canada's pre-eminent solid state physicists and educators.

Dr. Roman Maev.

Early life and education

Roman Maev was born in 1945 in Moscow, in the former USSR. He received his M.Sc. degree in Theoretical Nuclear Physics from the Moscow Physical Engineering Institute in 1969, where he was bestowed with the so-called Red Diploma, indicating that his marks never fell below the level of an A. In 1972, he received a Prize in Physics from the National Young Scientist Society. In 1973, he received his Ph.D. in the Theory of Semiconductors from the P.N. Lebedev Physical Institute of the USSR Academy of Sciences.[2] His doctoral research was prepared at the Laboratory of Quantum Radiophysics of that Institute, lead by Academician Nikolay Basov, a Nobel Prize winner best known for his work in quantum electronics which led to the development of lasers and masers.

Scientific and academic career

In 1978, Dr. Maev was appointed as Head of the Laboratory for Biophysical Introscopy at the Institute of Chemical Physics of the USSR Academy of Sciences, and in 1984, he began to serve as Acting Chairholder of the Biomedical Physics Chair at the Moscow Institute of Physics and Technology. In 1987, he established and became Founding Director of the Acoustic Microscopy Center at the USSR Academy of Sciences. In 1990, he received a Fellowship from the Gore-Chernomirdin Commission, and as a result was invited by the US government to take a special series of courses at Harvard Business School, where he stidied for several months, through its Scientific Business Management Fund. From 1990 to 1995, Dr. Maev was deeply involved as one of the leaders in the Russian Governmental Program in Technology Transfer. During that period, Maev, as one of the directors of that Program, together with his team of experts, successfully transferred about a dozen high-tech technologies with an average value of $20–250 million, primarily with Europe.

In the mid-1990s, Dr. Maev moved to Canada through an intergovernmental exchange program. He holds both Russian and Canadian citizenship. In 1995, Roman Maev was appointed a Full Faculty Professor in the Department of Physics at the University of Windsor in Ontario, Canada,[3] and in 2002, he became a Chairholder of the DaimlerChrysler Industrial Research Chair,[4] receiving the title of Distinguished University Professor. Since beginning his research activity in Canada, which has been ongoing since 1995, Dr. Maev has received support from various industrial partners and government agencies and grants totalling over $16 million.

As a worldwide authority in high resolution acoustic imaging and advanced material characterization, Dr. Maev, during his long research career, has been appointed an Adjunct Professor at Oxford University, Johns Hopkins University, McGill University, and the University of Michigan, in addition to being a member of the Brockhouse Materials Research Institute at McMaster University. He has held Visiting Professorships at the NIST, Rochester University, the University of Illinois at Urbana-Champaign, the University of California, Technische Universität München, Bundeswehr University, Kyoto University, Hefei University, Université Paris VI, Aberdeen University, and Università di Palermo. He is currently Adjunct Professor at the University of Toronto, and the Wroclaw Polytechnic University.[5]

In 2002, Roman Maev was bestowed a D.Sc. Degree in Physical-Mathematical Sciences with Special Recognition (in the Methods of Acoustic Microscopy of Investigation of Microstructure, Physical and Chemical Properties of Materials) from the Russian Academy of Sciences for the body of work he had so far produced in his career, and in 2005 he received a Full Professorship in Physics from the Government of the Russian Federation.

Dr. Maev currently serves as a member of the Editorial Advisory Board of the Journal of Research in Nondestructive Evaluation and as an Associate Editor of the IEEE's Transaction in Ultrasonics, Ferroelectrics and Frequency Control.

Awards

Roman Maev has won many awards for his innovations, research discoveries, and inventions. In recognition of his contribution to the development of ultrasound technique, he was awarded the Pioneer Award by the American Institute of Ultrasound in Medicine in 1988. In 1989, he was awarded the Centenary Ernst Abbe Medal from the World Microscopical Society. In 2001, he received a Letter of Recognition for Research Excellence from the Deputy Prime Minister of Canada. In 2001, 2002, and 2006, he received Awards for Outstanding Research and Development from the DaimlerChrysler Corporation. In 2001, 2004, 2005, 2007, and 2010, the University of Windsor bestowed him with Awards in Recognition of Research and Scholarship Excellence. In 2003, Dr. Maev received the Canada Innovation Summit Award in Recognition of Contribution to New Knowledge and Technical Innovation. In 2007, he received the Canadian Association of Physicists Gold Medal for Outstanding Achievement in Industrial and Applied Physics.[6] In 2007, he received the Ontario Premier's Catalyst Award for the Start Up Company with the Best Innovation. In the same year, he was awarded the Canadian Association of Physicists Medal for Outstanding Achievement in Industrial and Applied Physics. In 2009, he was elected as a Fellow of IEEE[7] and in 2010 he was elected as a Full Member of the Russian A. M. Prokhorov Academy of Engineering Sciences.

Principal research contributions

In 1978, Roman Maev designed ultrasonic transducers using piezoelectric films of ZnO and CdS that could operate in the GHz range so as to enable high resolution acoustical microscopy. In 1980, he designed and constructed the first high resolution (500 MHz) transmission-mode acoustical microscope. In 1980, Maev began to collaborate with Leitz Inc., in Wetzlar, Germany to contribute to theory and design methodology that lead to the production of the world's first commercially available scanning acoustical microscope, the ELSAM, in 1983. From 1984 to 1990, along with his associates from his own Research School, he developed the theory to determine the amplitude of acoustic waves occurring in transmission-mode microscopy and derived new quantitative amplitude-based method for more accurate material characterisation. In 1989, Maev designed, built, and commercialized a new portable commercial receiving-mode acoustic microscope which was implemented in various research institutions in Russia, Ukraine, Latvia, China and Germany. Later, in 2001, he developed a novel portable hand-held high-frequency acoustic imaging system for the characterization of bulk structures of advanced materials, such as metals and alloys, ceramics, composites, polymers, and bio-polymers.

Over the course of his career, Roman Maev has established world renowned Research Schools, first in Russia, and later in Canada. He has mentored over 125 graduate students, many of whom have gone on to leadership positions in academic and the private sector worldwide. Currently, an additional 26 scholars and innovators are being supervised. Maev has also helped establish three Research Centres and one Research Institute. He has created four spin-off companies.

Publications and Patents

The diverse range of disciplines encompassed by Dr. Maev includes theoretical fundamentals of solid state physics, physical acoustics, experimental research in ultrasonic and nonlinear acoustical imaging, and the theory of propagation of waves through layered structures. As of 2010, he had published four monographs and over 340 research peer-reviewed papers. He holds over 25 patents.

International Activities

Dr. Maev is currently a member of the Canada-Russia Intergovernmental Economical Commission and also represents the Russian Corporation of Nanotechnologies in Canada. In 2008, Roman Maev was appointed an Honorary Counsel of the Russian Federation in Canada (Windsor, Ontario).[8]

Tessonics

In 2004, Dr. Maev became a founder of Tessonics Inc., a Canadian high-tech start-up company which develops and manufactures cutting-edge ultrasonic imaging technologies for industrial and medical applications. Since 2004, Tessonics has grown and established itself as a company with a global market. As of 2010, Tessonics successfully sells its high-tech products to over 16 countries, and its clients include the world's leading automakers and suppliers.[9]

The Institute for Diagnostic Imaging Research (IDIR)

In 2008, Dr. Maev became the Founding Director of The Institute for Diagnostic Imaging Research, a multi-disciplinary, collaborative research and innovation consortium. The Institute was formed in 2008 in conjunction with the Ontario Ministry of Research and Innovation, which provided an initial research investment of $5 million. The Institute uses applied physics to create innovative imaging research that is often commercialized through technology transfer initiatives that include the private sector and the development of spin-off companies.

Personal life

Roman Maev is married to Russian-Canadian physicist Dr. Elena Maeva. They reside in Windsor, Ontario with their daughter and son.

References

[1] IDIR, Windsor, Ontario, Canada, wwww.idirresearch.com
[2] Maev, Roman Gr. Acoustic Microscopy: Fundamentals and Applications
[3] University of Windsor, Windsor, Ontario
[4] http://www.nserc-crsng.gc.ca/Partners-Partenaires/Chairholders-TitulairesDeChaire/Chairholder-Titulaire_eng.asp?pid=346
[5] University of Toronto, Faculty of Applied Science and Engineering
[6] The Canadian Association of Physicists http://www.cap.ca/awards/press/2007-Maev.html
[7] The Institute of Electrical and Electronics Engineers
[8] Foreign Affairs and International Trade Canada, http://w01.international.gc.ca/Protocol-Protocole/pdf/DrsBook_2008_11_eng.pdf
[9] www.tessonics.com

Nikolay_Basov

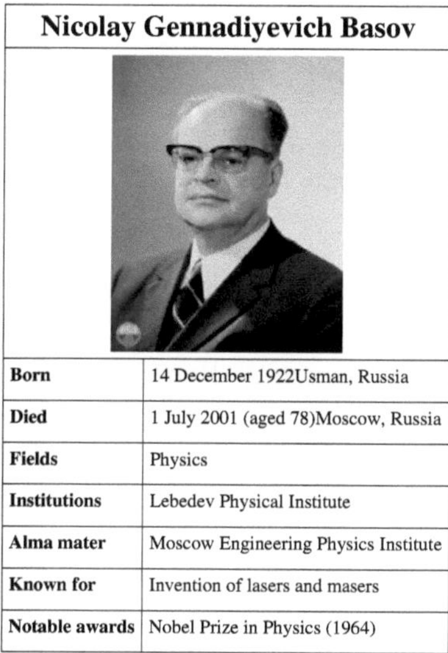

Nicolay Gennadiyevich Basov	
Born	14 December 1922Usman, Russia
Died	1 July 2001 (aged 78)Moscow, Russia
Fields	Physics
Institutions	Lebedev Physical Institute
Alma mater	Moscow Engineering Physics Institute
Known for	Invention of lasers and masers
Notable awards	Nobel Prize in Physics (1964)

Nikolay Gennadiyevich Basov (Russian: Никола́й Генна́диевич Ба́сов; 14 December 1922 − 1 July 2001) was a Soviet physicist and educator. For his fundamental work in the field of quantum electronics that led to the development of laser and maser, Basov shared the 1964 Nobel Prize in Physics with Alexander Prokhorov and Charles Hard Townes.[1]

Early life

Basov was born in the town Usman, now in Lipetsk Oblast in 1922.[2] He finished school in 1941 in Voronezh, and was later called for the military service at Kuibyshev Military Medical Academy. In 1943 he left academy and served in the Red Army[2] participating in the Second World War with the 1st Ukrainian Front.

Professional career

Basov graduated from Moscow Engineering Physics Institute (MEPI) in 1950. He then held a professorship at MEPI and also worked in the Lebedev Physical Institute (LPI), where defended a dissertation for the *Candidate of Sciences* degree (equivalent to PhD) in 1953 and a dissertation for the *Doctor of Sciences* degree in 1956. Basov was the Director of the LPI in 1973-1988. He was elected as a corresponding member of the USSR Academy of Sciences (Russian Academy of Sciences since 1991) in 1962 and an Full Member of the Academy in 1966. In 1967 he was elected a Member of the Presidium of the Academy (1967—1990), and since 1990 he was the councillor of the Presidium of the USSR Academy of Sciences. He was an honorary member of the International Academy of Science. He was the head of the laboratory of quantum radiophysics at the LPI until his death in 2001.[1]

Missile defense

Basov's contributions to the development of the laser and maser, which won him the Nobel Prize in 1964, led to new missile defense initiatives seeking to employ them.[3]

Politics

He entered politics in 1951 and a member of parliament of the Supreme Soviet in 1974.[2] Following U.S. President Ronald Reagan's speech on SDI in 1983, Basov signed a letter along with other Soviet scientists condemning the initiative, which was published in the New York Times.[4] In 1985 he declared the Soviet Union was capable of matching SDI proposals made by the U.S.[4]

Books

- N. G. Basov, K. A. Brueckner (Editor-in-Chief), S. W. Haan, C. Yamanaka. *Inertial Confinement Fusion*, 1992, Research Trends in Physics Series published by the American Institute of Physics Press (presently Springer [5], New York). ISBN 0-88318-925-9.
- V. Stefan and N. G. Basov (Editors). *Semiconductor Science and Technology*, Volume 1. Semiconductor Lasers. (Stefan University Press Series on Frontiers in Science and Technology) (Paperback), 1999. ISBN 1-889545-11-2.
- V. Stefan and N. G. Basov (Editors). *Semiconductor Science and Technology*, Volume 2: Quantum Dots and Quantum Wells. (Stefan University Press Series on Frontiers in Science and Technology) (Paperback), 1999. ISBN 1-889545-12-0.

Awards and honours

This article incorporates information from the equivalent article on the Russian Wikipedia.

- Lenin Prize (1959)
- Nobel Prize in Physics (1964, with the pioneering work done in the field of quantum electronics)
- Hero of Socialist Labour — twice (1969, 1982)
- Gold Medal of the Czechoslovak Academy of Sciences (1975)
- A. Volta Gold Medal (1977)
- Kalinga Prize (1986)
- USSR State Prize (1989)
- Lomonosov Grand Gold Medal, Moscow State University (1990)
- Order of Lenin - five times
- Order of Merit for the Fatherland, 2nd class
- Order of the Patriotic War, 2nd class

References

[1] "Basov Nikolay Gennadiyevich" (http://i-lasers.com/basov.html)

[2] "A century of Nobel Prizes recipients: chemistry, physics, and medicine" (http://books.google.com/books?id=3-G3vi5av28C), Francis Leroy. CRC Press, 2003. ISBN 0824708768, 9780824708764. p. 174-175

[3] "Soviet ballistic missile defense and the Western alliance" (http://books.google.com/books?id=fnqN-1UaNI4C), David Scott Yost. Harvard University Press, 1988. ISBN 0674826108, 9780674826106. p. 58

[4] "The strategic defence initiative: US policy and the Soviet Union" (http://books.google.com/books?id=J6cx2ZOixcYC), Mira Duric. Ashgate Publishing, Ltd., 2003. ISBN 0754637336, 9780754637332. p. 43-45

[5] http://www.springer-sbm.de/index.php?id=121&L=0

See also

- Excimer laser

External links

- Basov's grave (http://novodevichye.com/basov/)
- Detailed biography (http://www.nature.ru/db/msg.html?mid=1165378&s=120000000) (Russian)
- Biography (http://nobelprize.org/physics/laureates/1964/basov-bio.html) from the *Nobel Foundation*.
- Oral History interview transcript with Nikolay Basov 14 September 1984, American Institute of Physics, Niels Bohr Library and Archives (http://www.aip.org/history/ohilist/4495.html)

Nobel_Prize_in_Physics

The Nobel Prize in Physics	
Awarded for	Outstanding contributions in Physics
Presented by	Royal Swedish Academy of Sciences
Country	Sweden
First awarded	1901
Official website	nobelprize.org [1]

The **Nobel Prize in Physics** (Swedish: *Nobelpriset i fysik*) is awarded once a year by the Royal Swedish Academy of Sciences. It is one of the five Nobel Prizes established by the will of Alfred Nobel in 1895 and awarded since 1901; the others are the Nobel Prize in Chemistry, Nobel Prize in Literature, Nobel Peace Prize, and Nobel Prize in Physiology or Medicine. The first Nobel Prize in Physics was awarded to Wilhelm Conrad Röntgen, a German, "in recognition of the extraordinary services he has rendered by the discovery of the remarkable rays (or x-rays)." This award is administered by the Nobel Foundation and widely regarded as the most prestigious award that a scientist can receive in physics. It is presented in Stockholm at an annual ceremony on December 10, the anniversary of Nobel's death.

Wilhelm Röntgen (1845–1923), the first recipient of the Nobel Prize in Physics.

Background

Alfred Nobel requested in his last will and testament that his money be used to create a series of prizes for those who confer the "greatest benefit on mankind" in physics, chemistry, peace, physiology or medicine, and literature.[2] [3] Though Nobel wrote several wills during his lifetime, the last was written a little over a year before he died, and signed at the Swedish-Norwegian Club in Paris on 27 November 1895.[4] [5] Nobel bequeathed 94% of his total assets, 31 million Swedish *kronor* (US$186 million in 2008), to establish and endow the five Nobel Prizes.[6] Due to the level of skepticism surrounding the will it was not until April 26, 1897 that it was approved by the Storting (the Norwegian Parliament).[7] [8] The executors of his will were Ragnar Sohlman and Rudolf Lilljequist, who formed the Nobel Foundation to take care of Nobel's fortune and organise the prizes.

The members of the Norwegian Nobel Committee who were to award the Peace Prize were appointed shortly after the will was approved. The prize-awarding organisations followed: the Karolinska Institutet on June 7, the Swedish Academy on June 9, and the Royal Swedish Academy of Sciences on June 11.[9] [10] The Nobel Foundation then reached an agreement on guidelines for how the Nobel Prize should be awarded. In 1900, the Nobel Foundation's newly created statutes were promulgated by King Oscar II.[8] [11] [12] According to Nobel's will, The Royal Swedish Academy of sciences were to award the Prize in Physics.[12]

Nomination and selection

A maximum of three Nobel laureates and two different works may be selected for the Nobel Prize in Physics.[13] Compared with some other Nobel Prizes, the nomination and selection process for the prize in Physics is long and rigorous. This is a key reason it has grown in importance over the years to become the most important prize in Physics.[14]

The Nobel laureates are selected by the Nobel Committee for Physics, a Nobel Committee that consists of five members elected by The Royal Swedish Academy of Sciences. In the first stage, several thousand people are asked to nominate candidates. These names are scrutinized and discussed by experts until the choice is made.

Forms are sent to about three thousand individuals to invite them to submit nominations. The names of the nominees are never publicly announced, and neither are they told that they have been considered for the prize. Nomination records are sealed for fifty years. In practice, some nominees do become known. It is also common for publicists to make such a claim, founded or not.

The nominations are screened by committee, and a list is produced of approximately two hundred preliminary candidates. This list is forwarded to selected experts in the field. They narrow it down to approximately fifteen names. The committee submits a report with recommendations to the appropriate institution.

While posthumous nominations are not permitted, awards can be made if the individual died in the months between the decision of the prize committee (typically in October) and the ceremony in December. Prior to 1974, posthumous awards were permitted if the recipient had died after being nominated.[15]

The rules for ihe Nobel Prize in Physics require that the significance of achievements being recognized has been "tested by time." In practice it means that the lag between the discovery and the award is typically on the order of 20 years and can be much longer. For example, half of the 1983 Nobel Prize in Physics was awarded to Subrahmanyan Chandrasekhar for his work on stellar structure and evolution that was done during the 1930s. As a downside of this approach, not all scientists live long enough for their work to be recognized. Some important scientific discoveries are never considered for a prize, as the discoverers may have died by the time the impact of their work is appreciated.[16] [17] [18]

Prizes

A Physics Nobel Prize laureate earns a gold medal, a diploma bearing a citation, and a sum of money.[19] The amount of money awarded depends on the income of the Nobel Foundation that year.[20] If a prize is awarded to more than one laureate, the money is either split evenly among them or, for three laureates, it may be divided into a half and two quarters.[21] If a prize is awarded jointly to two or more laureates, the money is split among them.[21]

Medals

The Nobel Prize medals, minted by Myntverket[22] in Sweden and the Mint of Norway since 1902, are registered trademarks of the Nobel Foundation. Each medal has an image of Alfred Nobel in left profile on the obverse (front side of the medal). The Nobel Prize medals for Physics, Chemistry, Physiology or Medicine, and Literature have identical obverses, showing the image of Alfred Nobel and the years of his birth and death (1833–1896). Nobel's portrait also appears on the obverse of the Nobel Peace Prize medal and the Medal for the Prize in Economics, but with a slightly different design.[23] [24] The image on the reverse of a medal varies according to the institution awarding the prize. The reverse sides of the Nobel Prize medals for Chemistry and Physics share the same design.[25]

Diplomas

Nobel laureates receive a diploma directly from the hands of the King of Sweden. Each diploma is uniquely designed by the prize-awarding institutions for the laureate that receives it.[26] The diploma contains a picture and text which states the name of the laureate and normally a citation of why they received the prize.[26]

Award money

The laureate is also given a sum of money when they receive the Nobel Prize in the form of a document confirming the amount awarded; in 2009, the monetary award was 10 million SEK (US$1.4 million).[20] The amount may differ depending on how much money the Nobel Foundation can award that year. If there are two winners in a particular category, the award grant is divided equally between the recipients. If there are three, the awarding committee has the option of dividing the grant equally, or awarding one-half to one recipient and one-quarter to each of the others.[27] [28] [29] [30]

Ceremony

The committee and institution serving as the selection board for the prize typically announce the names of the laureates in October. The prize is then awarded at formal ceremonies held annually on 10 December, the anniversary of Nobel's death. "The highlight of the Nobel Prize Award Ceremony in Stockholm is when each Nobel Laureate steps forward to receive the prize from the hands of His Majesty the King of Sweden. ... Under the eyes of a watching world, the Nobel Laureate receives three things: a diploma, a medal and a document confirming the prize amount" ("What the Nobel Laureates Receive").

The Nobel Banquet is held every year in Stockholm City Hall in connection with the Nobel Prize.[31]

See also

- Sakurai Prize, presented by the American Physical Society
- List of Nobel laureates

Notes

[1] http://nobelprize.org

[2] "History – Historic Figures: Alfred Nobel (1833–1896)" (http://www.bbc.co.uk/history/historic_figures/nobel_alfred.shtml). BBC. . Retrieved 2010-01-15.

[3] "Guide to Nobel Prize" (http://www.britannica.com/nobelprize/article-9056008). Britannica.com. . Retrieved 2010-01-15.

[4] Ragnar Sohlman: 1983, Page 7

[5] von Euler, U.S. (6 June 1981). "The Nobel Foundation and its Role for Modern Day Science" (http://resources.metapress.com/pdf-preview.axd?code=xu7j67w616m06488&size=largest) (PDF). Die Naturwissenschaften (Springer-Verlag). . Retrieved 21 January 2010.

[6] "The Will of Alfred Nobel" (http://nobelprize.org/alfred_nobel/will/index.html), nobelprize.org. Retrieved 6 November 2007.

[7] "The Nobel Foundation – History" (http://nobelprize.org/nobelfoundation/history/lemmel/index.html). Nobelprize.org. . Retrieved 2010-01-15.

[8] Agneta Wallin Levinovitz: 2001, Page 13

[9] "Nobel Prize History —" (http://www.infoplease.com/spot/nobel-prize-history.html). Infoplease.com. 1999-10-13. . Retrieved 2010-01-15.

[10] Encyclopædia Britannica. "Nobel Foundation (Scandinavian organisation) – Britannica Online Encyclopedia" (http://www.britannica.com/EBchecked/topic/416852/Nobel-Foundation). Britannica.com. . Retrieved 2010-01-15.

[11] AFP, "Alfred Nobel's last will and testament" (http://www.thelocal.se/14776/20091005/), The Local(5 October 2009): accessed 20 January 2010.

[12] "Nobel Prize (http://www.britannica.com/EBchecked/topic/416856/Nobel-Prize)" (2007), in Encyclopædia Britannica, accessed 15 January 2009, from Encyclopædia Britannica Online:

    After Nobel's death, the Nobel Foundation was set up to carry out the provisions of his will and to administer his funds. In his will, he had stipulated that four different institutions—three Swedish and one Norwegian—should award the prizes. From Stockholm, the Royal Swedish Academy of Sciences confers the prizes for physics, chemistry, and economics, the Karolinska Institute confers the prize for physiology or medicine, and the Swedish Academy confers the prize for literature. The Norwegian Nobel Committee based in Oslo confers the prize for peace. The Nobel Foundation is the legal owner and functional administrator of the funds and serves as the joint administrative body of the prize-awarding institutions, but it is not concerned with the prize deliberations or decisions, which rest exclusively with the four institutions.

[13] "What the Nobel Laureates Receive" (http://nobelprize.org/award_ceremonies/prize.html). Retrieved November 1, 2007. Archived (http://web.archive.org/20071030033639/http://nobelprize.org/award_ceremonies/prize.html) October 30, 2007 at the Wayback Machine

[14] "The Nobel Prize Selection Process" (http://www.britannica.com/eb/art-18050), *Encyclopædia Britannica*, accessed November 5, 2007 (Flowchart).

[15] FAQ nobelprize.org (http://nobelprize.org/contact/faq/index.html)

[16] "web-041003.dvi" (http://nobelprize.org/nobel_prizes/economics/laureates/2004/ecoadv.pdf) (PDF). . Retrieved 2010-02-05.

[17] Gingras, Yves; Wallace, Matthew L. (2009). "Why it has become more difficult to predict Nobel Prize winners: A bibliometric analysis of nominees and winners of the chemistry and physics prizes (1901–2007)". *Scientometrics* **82** (2): 401. doi:10.1007/s11192-009-0035-9.

[18] *Nature Chemistry*. Bibcode 2009NatCh...1..509.. doi:10.1038/nchem.372.

[19] Tom Rivers (2009-12-10). "2009 Nobel Laureates Receive Their Honors | Europe| English" (http://www1.voanews.com/english/news/europe/2009-Nobel-Laureates-Receive-Their-Honors-78989292.html). .voanews.com. . Retrieved 2010-01-15.

[20] "The Nobel Prize Amounts" (http://nobelprize.org/nobel_prizes/amounts.html). Nobelprize.org. . Retrieved 2010-01-15.

[21] "Nobel Prize – Prizes" (http://www.britannica.com/EBchecked/topic/416856/Nobel-Prize/93434/The-prizes) (2007), in *Encyclopædia Britannica*, accessed 15 January 2009, from *Encyclopædia Britannica Online*:

Each Nobel Prize consists of a gold medal, a diploma bearing a citation, and a sum of money, the amount of which depends on the income of the Nobel Foundation. (A sum of $1,300,000 accompanied each prize in 2005.) A Nobel Prize is either given entirely to one person, divided equally between two persons, or shared by three persons. In the latter case, each of the three persons can receive a one-third share of the prize or two together can receive a one-half share.

[22] "Medalj – ett traditionellt hantverk" (http://www.myntverket.se/products.asp?lang=sv&page=3) (in Swedish). Myntverket. . Retrieved 2007-12-15.

[23] "The Nobel Prize for Peace" (http://digitalcollections.library.oregonstate.edu/cdm4/item_viewer.php?CISOROOT=/pawardsmedals&CISOPTR=47), "Linus Pauling: Awards, Honors, and Medals", *Linus Pauling and The Nature of the Chemical Bond: A Documentary History*, the Valley Library, Oregon State University. Retrieved 7 December 2007.

[24] "The Nobel Medals" (http://www.ceptualinstitute.com/galleria/awards/nobel/nobelmedals.html). Ceptualinstitute.com. . Retrieved 2010-01-15.

[25] "Nobel Prize for Chemistry. Front and back images of the medal. 1954" (http://osulibrary.oregonstate.edu/specialcollections/coll/pauling/bond/pictures/nobel-chemistry-medal.html), "Source: Photo by Eric Arnold. Ava Helen and Linus Pauling Papers. Honors and Awards, 1954h2.1", "All Documents and Media: Pictures and Illustrations", *Linus Pauling and The Nature of the Chemical Bond: A Documentary History*, the Valley Library, Oregon State University. Retrieved 7 December 2007.

[26] "The Nobel Prize Diplomas" (http://nobelprize.org/nobel_prizes/diplomas/). Nobelprize.org. . Retrieved 2010-01-15.

[27] Sample, Ian (2009-10-05). "Nobel prize for medicine shared by scientists for work on ageing and cancer | Science | guardian.co.uk" (http://www.guardian.co.uk/science/2009/oct/05/nobel-prize-medicine-2009-award). London: Guardian. . Retrieved 2010-01-15.

[28] Ian Sample, Science correspondent (2008-10-07). "Three share Nobel prize for physics | Science | guardian.co.uk" (http://www.guardian.co.uk/science/2008/oct/07/physics.nobel). London: Guardian. . Retrieved 2010-02-10.

[29] David Landes. "Americans claim Nobel economics prize – The Local" (http://www.thelocal.se/22604/20091012/). Thelocal.se. . Retrieved 2010-01-15.

[30] "The 2009 Nobel Prize in Physics - Press Release" (http://nobelprize.org/nobel_prizes/physics/laureates/2009/press.html). Nobelprize.org. 2009-10-06. . Retrieved 2010-02-10.

[31] Nobel Prize Foundation Website (http://nobelprize.org/award_ceremonies/banquet/menus/soderlind/index.html)

References

- Friedman, Robert Marc (2001). *The Politics of Excellence: Behind the Nobel Prize in Science*. New York & Stuttgart: VHPS (Times Books). ISBN 0716731037, ISBN 978-0716731030.
- Gill, Mohammad (March 10, 2005). "Prize and Prejudice" (http://www.chowk.com/articles/8836). *Chowk* magazine.
- Hillebrand, Claus D. (June 2002). "Nobel century: a biographical analysis of physics laureates" (http://www. ingentaconnect.com/content/maney/isr/2002/00000027/00000002/art00003). *Interdisciplinary Science Reviews* 27.2: 87-93.
- Schmidhuber, Jürgen (2010). Evolution of National Nobel Prize Shares in the 20th Century (http://www.idsia. ch/~juergen/nobelshare.html) at arXiv:1009.2634v1 (http://arxiv.org/abs/1009.2634) with graphics: National Physics Nobel Prize shares 1901-2009 by citizenship at the time of the award (http://www.idsia.ch/~juergen/ phys.html) and by country of birth (http://www.idsia.ch/~juergen/physnat.html).
- Lemmel, Birgitta. "The Nobel Prize Medals and the Medal for the Prize in Economics" (http://nobelprize.org/ nobel_prizes/medals/). *nobelprize.org*. Copyright © The Nobel Foundation 2006. (An article on the history of the design of the medals.)
- "What the Nobel Laureates Receive" (http://nobelprize.org/award_ceremonies/prize.html). *nobelprize.org*. Copyright © Nobel Web AB 2007.

External links

- "All Nobel Laureates in Physics" (http://nobelprize.org/nobel_prizes/physics/laureates/) - Index webpage on the official site of the Nobel Foundation.
- "The Nobel Prize Award Ceremonies" (http://nobelprize.org/award_ceremonies/) – Official hyperlinked webpage of the Nobel Foundation.
- "The Nobel Prize in Physics" (http://www.nobelprize.org/nobel_prizes/physics/) - Official webpage of the Nobel Foundation.
- "The Nobel Prize Medal for Physics and Chemistry" (http://nobelprize.org/nobel_prizes/physics/medal.html) – Official webpage of the Nobel Foundation.

Alexander_Prokhorov

<table>
<tr><td colspan="2" align="center">Alexander Prokhorov</td></tr>
<tr><td>Born</td><td>11 July 1916Atherton, Queensland, Australia</td></tr>
<tr><td>Died</td><td>8 January 2002 (aged 85)Moscow, Russia</td></tr>
<tr><td>Nationality</td><td>Soviet
Russian</td></tr>
<tr><td>Fields</td><td>Physics</td></tr>
<tr><td>Known for</td><td>Lasers and masers</td></tr>
<tr><td>Notable awards</td><td>1964 Nobel Prize in Physics</td></tr>
</table>

Alexander Mikhaylovich Prokhorov (Russian: Алекса́ндр Миха́йлович Про́хоров) (11 July 1916 – 8 January 2002) was a Soviet physicist known for his pioneering research on lasers and masers for which he shared the Nobel Prize in Physics in 1964 with Charles Hard Townes and Nikolay Basov.

Early life

Prokhorov was born in 1916 in Atherton, Queensland, to a family of Russian revolutionaries who emigrated from Russia to escape repression by the tsarist government. In 1923, after the October Revolution, they returned to Russia. In 1934, Prokhorov entered the Saint Petersburg State University to study physics. He graduated with honors in 1939 and moved to Moscow to work at the Lebedev Physical Institute, in the oscillations laboratory headed by academician N. D. Papaleksi. His research there was devoted to propagation of radio waves in the ionosphere. At the onset of World War II in the Soviet Union, in June 1941, he joined the Red Army. During World War II, Prokhorov fought in the infantry, was wounded twice in battles, and was awarded three medals, including the Medal of Valour in 1946.[1] He was demobilized in 1944 and returned to the Lebedev Institute where in 1946 defended his Ph.D. thesis on "Theory of Stabilization of Frequency of a Tube Oscillator in the Theory of a Small Parameter".[2] [3] [4]

Research

In 1947, Prokhorov started working on coherent radiation emitted by electrons orbiting in a cyclic particle accelerator called a synchrotron. He demonstrated that the emission is mostly concentrated in the microwave spectral range. His results became the basis of his habilitation on "Coherent Radiation of Electrons in the Synchrotron Accelerator", defended in 1951. By 1950, Prokhorov was assistant chief of the oscillation laboratory. Around that time, he formed a group of young scientists to work on radiospectroscopy of molecular rotations and vibrations, and later on quantum electronics. The group focused on a special class of molecules which have three (non-degenerate) moments of inertia. The research was conducted both on experiment and theory. In 1954, Prokhorov became head of the laboratory. Together with Nikolay Basov he developed theoretical grounds for creation of a molecular oscillator and constructed such an oscillator based on ammonia. They also proposed a method for the production of population inversion using inhomogeneous electric and magnetic fields. Their results were first presented at a national conference in 1952, but not published until 1954–1955;[2] [4]

In 1955, Prokhorov started his research in the field of electron paramagnetic resonance (EPR). He focused on relaxation times of ions of the iron group elements in a lattice of aluminium oxide, but also investigated other, "non-optical", topics, such as magnetic phase transitions in DPPH.[5] In 1957, while studying ruby, a chromium-doped variation of aluminium oxide, he came upon the idea of using this material as an active medium of a laser. As a new type of laser resonator, he proposed, in 1958, an "open type" cavity design, which is widely used today. In 1963, together with A. S. Selivanenko, he suggested a laser using two-quantum transitions. For his pioneering work on lasers and masers, in 1964, he received the Nobel Prize in Physics shared with Nikolay Basov and Charles Hard Townes.[2] [4]

Posts and awards

In 1959, Prokhorov became a professor at Moscow State University – the most prestigious university in the Soviet Union; the same year, he was awarded the Lenin Prize. In 1960, he became a member of the Russian Academy of Sciences and elected Academician in 1966. In 1967, he was awarded his first Order of Lenin (he received five of them during life, in 1967, 1969, 1975, 1981 and 1986). In 1968, he became vice-director of the Lebedev Institute and in 1971 took the position of Head of Laboratory of another prestigious Soviet institution, the Moscow Institute of Physics and Technology. In the same year, he was elected a member of the American Academy of Arts and Sciences.[3] Between 1982 and 1998, Prokhorov served as acting director of the General Physics Institute of the Russian Academy of Sciences, and after 1998 as honorary director. After his death in 2002, the institute was renamed the A. M. Prokhorov General Physics Institute of the Russian Academy of Sciences.[3] [4]

In 1969, Prokhorov became a Hero of Socialist Labour, the highest degree of distinction in the Soviet Union for achievements in national economy and culture. He received the second such award in 1986.[4] Starting in 1969, he was the chief editor of the Great Soviet Encyclopedia. He was awarded the Frederic Ives Medal, the highest distinction of the Optical Society of America (OSA), in 2000[6] and became an Honorary OSA Member in 2001.[7] The same year, he was awarded the Demidov Prize.[8]

Politics

Prokhorov became member of the Communist Party in 1950. In 1983, together with three other academicians – Andrey Tychonoff, Anatoly Dorodnitsyn and Georgy Skryabin – he signed the famous open letter[9] denouncing Andrey Sakharov's article[10] in the *Foreign Affairs*.

Family

Both of Prokhorov's parents died during World War II. Prokhorov married geographer Galina Shelepina in 1941, and they had a son, Kiril, born in 1945. Following his father, Kiril Prokhorov became a physicist in the field of optics and is currently leading a laser-related laboratory at the A. M. Prokhorov General Physics Institute.[1] [11]

Honours and awards

This article incorporates information from the equivalent article on the Russian Wikipedia.

- Mandelstam Prize (1948)
- Lenin Prize (1959)
- Five Orders of Lenin (including 11 May 1981)
- Order of the Patriotic War, 1st class (1985)
- Nobel Prize in Physics (1964)
- Hero of Socialist Labour, twice (1969, 1986)
- Medal of Valour
- USSR State Prize (1980)
- Order of Merit for the Fatherland, 2nd class (1996)
- State Prize of the Russian Federation (1998)
- Medal Frederick Ayvesa (2000)
- Demidov Prize (2001)
- Lomonosov Gold Medal (Moscow State University, 1987)
- Award of the Council of Ministers
- State Prize of the Russian Federation in science and technology (2003, posthumously) for the development of scientific and technological foundations of metrological support of measurements of length in the microwave and nanometer ranges and their application in microelectronics and nanotechnology
- Foreign Member of the Czechoslovak Academy of Sciences (1982)
- Medal "In commemoration of the 100th anniversary of the birth of Vladimir Ilyich Lenin"
- Medal for the Victory Over Germany in the Great Patriotic War 1941–1945
- 20 years of victory
- 30 years of victory
- Jubilee Medal "Forty Years of Victory in Great Patriotic War of 1941-1945."
- Valiant Labour in the Great Patriotic War of 1941–1945 Medal
- Veteran of Labour Medal
- Medal "50 Years of the USSR Armed Forces"
- 800th Anniversary of Moscow Medal
- 850th Anniversary of Moscow Medal

Books

- A. M. Prokhorov (Editor in Chief), J. M. Buzzi, P. Sprangle, K. Wille. *Coherent Radiation Generation and Particle Acceleration*, 1992, ISBN 0-88318-926-7. *Research Trends in Physics* series published by the American Institute of Physics Press (presently Springer [5], New York)
- V. Stefan and A. M. Prokhorov (Editors) *Diamond Science and Technology Vol 1: Laser Diamond Interaction. Plasma Diamond Reactors* (Stefan University Press Series on Frontiers in Science and Technology) 1999 ISBN 1-889545-23-6.
- V. Stefan and A. M. Prokhorov (Editors). *Diamond Science and Technology Vol 2* (Stefan University Press Series on Frontiers in Science and Technology) 1999 ISBN 1-889545-24-4.

References

[1] Основные даты жизни и деятельности академика А.М. Прохорова (http://www.gpi.ru/memory/amp.php) (in Russian)

[2] Aleksandr M. Prokhorov, The Nobel Prize in Physics 1964 (http://nobelprize.org/nobel_prizes/physics/laureates/1964/prokhorov-bio. html)

[3] Прохоров Александр Михайлович (http://bse.sci-lib.com/article093736.html) in Great Soviet Encyclopedia (in Russian)

[4] Прохоров Александр Михайлович (http://www.warheroes.ru/hero/hero.asp?Hero_id=10290) at warheroes.ru (in Russian)

[5] A. M. Prokhorov and V.B. Fedorov, Soviet Physics JETP 16 (1963) 1489.

[6] Frederic Ives Medal (http://www.osa.org/aboutosa/awards/osaawards/awardsdesc/ivesquinn/)

[7] OSA Honorary Members (http://www.osa.org/aboutosa/awards/honorarymembers/)

[8] Сост. И. Г. Бебих, Г. Н. Михайлова, А. В. Троицкий (2004) (in Russian). *Александр Михайлович Прохоров, 1916-2002 (Материалы к биобиблиогр. ученых) 2-е изд., доп.* (http://www.naukaran.ru/sb/2003_3-4/17.shtml). М.: Наука. p. 442. ISBN 5020331767. .

[9] Когда теряют честь и совесть (http://www.ihst.ru/projects/sohist/material/press/sakharov/83.htm)(in Russian)

[10] Hawks Get Reluctant Ally in Sakharov (http://news.google.com/newspapers?id=9EMwAAAAIBAJ&sjid=h6UFAAAAIBAJ& pg=1164,1223149&dq=foreign+affairs+sakharov&hl=en)

[11] Кирилл Александрович Прохоров (http://www.gpi.ru/staff_s.php?eng=0&id=44)

External links

- Prokhorov's role in the invention of lasers and masers (http://i-lasers.com)

Quantum_optics

Quantum optics is a field of research in physics, dealing with the application of quantum mechanics to phenomena involving light and its interactions with matter.

History of quantum optics

Light is made up of particles called photons and hence inherently is "grainy" (quantized). Quantum optics is the study of the nature and effects of light as quantized photons. The first indication that light might be quantized came from Max Planck in 1899 when he correctly modeled blackbody radiation. By assuming blackbody radiation is quantized, Bohr showed that the atoms were also quantized, in the sense that they could only emit discrete amounts of energy. The understanding of the interaction between light and matter following these developments not only formed the basis of quantum optics but was also crucial for the development of quantum mechanics as a whole. However, the subfields of quantum mechanics dealing with matter-light interaction were principally regarded as research into matter rather than into light; hence one rather spoke of atom physics and quantum electronics in 1960. Laser science—i.e., research into principles, design and application of these devices—became an important field, and the quantum mechanics underlying the laser's principles was studied now with more emphasis on the properties of light, and the name *quantum optics* became customary.

As laser science needed good theoretical foundations, and also because research into these soon proved very fruitful, interest in quantum optics rose. Following the work of Dirac in quantum field theory, George Sudarshan, Roy J. Glauber, and Leonard Mandel applied quantum theory to the electromagnetic field in the 1950s and 1960s to gain a more detailed understanding of photodetection and the statistics of light (see degree of coherence). This led to the introduction of the coherent state as a quantum description of laser light and the realization that some states of light could not be described with classical waves. In 1977, Kimble et al. demonstrated the first source of light which required a quantum description: a single atom that emitted one photon at a time. This was the first conclusive evidence that light was made up of photons. Another quantum state of light with certain advantages over any classical state, squeezed light, was soon proposed. At the same time, development of short and ultrashort laser pulses—created by Q switching and modelocking techniques—opened the way to the study of unimaginably fast ("ultrafast") processes. Applications for solid state research (e.g. Raman spectroscopy) were found, and mechanical forces of light on matter were studied. The latter led to levitating and positioning clouds of atoms or even small biological samples in an optical trap or optical tweezers by laser beam. This, along with Doppler cooling was the crucial technology needed to achieve the celebrated Bose-Einstein condensation.

Other remarkable results are the demonstration of quantum entanglement, quantum teleportation, and (recently, in 1995) quantum logic gates. The latter are of much interest in quantum information theory, a subject which partly emerged from quantum optics, partly from theoretical computer science.

Today's fields of interest among quantum optics researchers include parametric down-conversion, parametric oscillation, even shorter (attosecond) light pulses, use of quantum optics for quantum information, manipulation of single atoms, Bose-Einstein condensates, their application, and how to manipulate them (a sub-field often called atom optics), coherent perfect absorbers, and much more.

Research into quantum optics that aims to bring photons into use for information transfer and computation is now often called photonics to emphasize the claim that photons and photonics will take the role that electrons and electronics now have.

Concepts of quantum optics

According to quantum theory, light may be considered not only as an electro-magnetic wave but also as a "stream" of particles called photons which travel with c, the vacuum speed of light. These particles should not be considered to be classical billiard balls, but as quantum mechanical particles described by a wavefunction spread over a finite region.

Each particle carries one quantum of energy equal to hf, where h is Planck's constant and f is the frequency of the light. The postulation of the quantization of light by Max Planck in 1899 and the discovery of the general validity of this idea in Albert Einstein's 1905 explanation of the photoelectric effect soon led physicists to realize the possibility of population inversion and the possibility of the laser.

This kind of use of statistical mechanics is the fundament of most concepts of quantum optics: Light is described in terms of field operators for creation and annihilation of photons—i.e. in the language of quantum electrodynamics.

A frequently encountered state of the light field is the coherent state as introduced by George Sudarshan in 1963. This state, which can be used to approximately describe the output of a single-frequency laser well above the laser threshold, exhibits Poissonian photon number statistics. Via certain nonlinear interactions, a coherent state can be transformed into a squeezed coherent state, which can exhibit super- or sub- Poissonean photon statistics. Such light is called squeezed light. Other important quantum aspects are related to correlations of photon statistics between different beams. For example, parametric nonlinear processes can generate so-called twin beams, where ideally each photon of one beam is associated with a photon in the other beam.

Atoms are considered as quantum mechanical oscillators with a discrete energy spectrum with the transitions between the energy eigenstates being driven by the absorption or emission of light according to Einstein's theory with the oscillator strength depending on the quantum numbers of the states.

For solid state matter one uses the energy band models of solid state physics. This is important as understanding how light is detected (typically by a solid-state device that absorbs it) is crucial for understanding experiments.

Quantum electronics

This term was used for the area of physics dealing with the effects of quantum mechanics on the behavior of electrons in matter, and their interactions with photons. It is today rarely considered a sub-field in its own right, as it has been absorbed by other fields. Solid state physics regularly takes quantum mechanics into account, and is usually concerned with electrons. Specific application to electronics is researched within semiconductor physics. The term also encompassed the basic processes of laser operation where photons are interacting with electrons: absorption, spontaneous emission, and stimulated emission. The term was mainly used between the 1950s and the 1970s. Today, the research output of this field is mainly used in quantum optics, especially for the part of it that draws not from atomic physics but from solid-state physics. Its usage overlapped Quantum Hall effect and Quantum cellular automata.

See also

- Optics
- Optical phase space
- Optical physics
- Nonclassical light

References

- L. Mandel, E. Wolf *Optical Coherence and Quantum Optics* (Cambridge 1995)
- D. F. Walls and G. J. Milburn *Quantum Optics* (Springer 1994)
- C. W. Gardiner and Peter Zoller, *Quantum Noise*, (Springer 2004).
- M. O. Scully and M. S. Zubairy *Quantum Optics* (Cambridge 1997)
- W. P. Schleich *Quantum Optics in Phase Space* (Wiley 2001)

External links

- An introduction to quantum optics of the light field [1]
- Encyclopedia of laser physics and technology [2], with content on quantum optics (particularly quantum noise in lasers), by Rüdiger Paschotta.
- Qwiki [3] - A quantum physics wiki devoted to providing technical resources for practicing quantum physicists.
- Quantiki [4] - a free-content WWW resource in quantum information science that anyone can edit.
- Various Quantum Optics Reports [5]

References

[1] http://gerdbreitenbach.de/gallery
[2] http://www.rp-photonics.com/encyclopedia.html
[3] http://qwiki.stanford.edu/wiki/Quantum_Optics
[4] http://www.quantiki.org/
[5] http://www.physics.drexel.edu/~tim/Decoherence/index.html

Physicist

A **physicist** is a scientist who does research in physics. Physicists study a wide range of physical phenomena in many branches of physics spanning all length scales: from sub-atomic particles of which all ordinary matter is made (particle physics) to the behavior of the material Universe as a whole (cosmology).

Etymology

The term "Physicist" was coined by English philosopher, priest, and historian of science William Whewell in 1840, to denote a cultivator of physics.[1]

Education

Most material a student encounters in the undergraduate physics curriculum is based on discoveries and insights of a century or more in the past. Alhazen's intromission theory of light was formulated in the 11th century; Newton's laws of motion and Newton's law of universal gravitation were formulated in the 17th century; Maxwell's equations, 19th century; and quantum mechanics, early 20th century. The undergraduate physics curriculum generally includes the following range of courses: chemistry, classical physics, kinematics, astronomy and astrophysics, physics laboratory, electricity and magnetism, thermodynamics, optics, modern physics, quantum physics, nuclear physics, particle physics, and solid state physics. Undergraduate physics students must also take extensive mathematics courses (calculus, differential equations, advanced calculus), and computer science and programming. Undergraduate physics students often perform research with faculty members.

Isaac Newton was a revolutionary figure in the development of modern physics as an exact science.

Many positions, especially in research, require a doctoral degree. At the Master's level and higher, students tend to specialize in a particular field. Fields of specialization include experimental and theoretical astrophysics, atomic physics, molecular physics, biophysics, chemical physics, medical physics, condensed matter physics, cosmology, geophysics, material science, nuclear physics, optics, particle physics, and plasma physics. Post-doctoral experience may be required for certain positions.

Albert Einstein developed the theory of general relativity.

Employment

The three major employers of career physicists are academic institutions, government laboratories, and private industries, with the largest employer being the last.[2] Many people who are trained as physicists, however, apply their skills to other activities, in particular to engineering, computing, and finance, often quite successfully. Some physicists take up additional careers where their knowledge of physics can be combined with further training in other disciplines, such as patent law in industry or private practice. In the United States, a majority of those in the private sections having a physics degree actually work outside the fields of physics, astronomy and engineering altogether.[3]

Nobel laureate Sir Joseph Rotblat has suggested that physicists going into employment in scientific research should honour a Hippocratic Oath for Scientists.

Honors and awards

The highest honor awarded to physicists is the Nobel Prize in Physics, awarded since 1901 by the Royal Swedish Academy of Sciences.

See also

- American Institute of Physics
- Engineering physics
- History of physics
- Institute of Physics (UK & Ireland)
- List of physicists
- Nobel Prize in physics
- Professional physicist

References

[1] http://www.etymonline.com/index.php?term=physics
[2] AIP Statistical Research Center. "Initial Employment Report, Fig. 7" (http://web.archive.org/web/20070113215906/http://www.aip. org/statistics/trends/highlite/emp/figure7.htm). Archived from the original (http://www.aip.org/statistics/trends/highlite/emp/figure7. htm) on January 13, 2007. . Retrieved August 21, 2006. Also relevant is: Institute of Physics. "Education Statistics, Graph 4.11" (http://www. iop.org/Our_Activities/Science_Policy/Statistics/Education Statistics/page_2680.html). . Retrieved August 21, 2006.
[3] AIP Statistical Research Center. "Initial Employment Report, Table 1" (http://www.aip.org/statistics/trends/highlite/emp/table1.htm). . Retrieved August 21, 2006.

Further reading

- Whitten, Barbara L.; Foster, Suzanne R.; Duncombe, Margaret L. (2003). "What works for women in physics?" (http://ptonline.aip.org/journals/doc/PHTOAD-ft/vol_56/iss_9/46_1.shtml). *Physics Today* **56** (9): 46. Bibcode 2003PhT....56i..46W. doi:10.1063/1.1620834.
- Kirby, Kate; Czujko, Roman; Mulvey, Patrick (2001). "The Physics Job Market: From Bear to Bull in a Decade" (http://ptonline.aip.org/journals/doc/PHTOAD-ft/vol_54/iss_4/36_1.shtml). *Physics Today* **54** (4): 36. Bibcode 2001PhT....54d..36K. doi:10.1063/1.1372112.
- Hermanowicz, Joseph C. (1998). *The Stars Are Not Enough: Scientists--Their Passions and Professions.* University of Chicago Press. ISBN 9780226327679.
- Hermanowicz, Joseph C. (2009). *Lives in Science: How Institutions Affect Academic Careers.* University of Chicago Press. ISBN 9780226327617.

External links

- Education and employment statistics (http://www.aip.org/statistics/) from the American Institute of Physics
- Occupational Outlook Handbook (http://www.bls.gov/oco/home.htm)
- Physicists and Astronomers (http://www.bls.gov/oco/ocos052.htm); US Department of Labor, Bureau of Labor Statistics
- PhysicistTv (http://www.physicisttv.com/)

Charles_Hard_Townes

Charles Hard Townes	
Born	July 28, 1915Greenville South Carolina
Residence	United States
Nationality	United States
Fields	Physics
Institutions	Berkeley Bell Labs Institute for Defense Analyses Columbia MIT
Alma mater	Furman University (B.S. & B.A.) Duke University (M.A.) Caltech (Ph.D.)
Doctoral advisor	William Smythe
Doctoral students	James P. Gordon Robert Boyd Ali Javan Raymond Y. Chiao
Known for	Inventing the Maser
Notable awards	Nobel Prize in Physics (1964) Templeton Prize (2005)

Charles Hard Townes (born July 28, 1915) is an American Nobel Prize-winning physicist and educator. Townes is known for his work on the theory and application of the maser, on which he got the fundamental patent, and other work in quantum electronics connected with both maser and laser devices. He shared the Nobel Prize in Physics in 1964 with Nikolay Basov and Alexander Prokhorov. The Japanese FM Towns computer and game console is named in his honour.

Biography

Townes was born in Greenville, South Carolina on July 28, 1915, the son of Henry Keith Townes, an attorney, and Ellen (Hard) Townes. He attended the Greenville public schools and then Furman University in Greenville, where he completed the requirements for the Bachelor of Science degree in Physics and the Bachelor of Arts degree in Modern Languages, graduating summa cum laude in 1935, at the age of 19. Physics had fascinated him since his first course in the subject during his sophomore year in college because of its "beautifully logical structure". He was also interested in natural history while at Furman, serving as curator of the museum, and working during the summers as collector for Furman's biology camp. In addition, he was busy with other activities, including the swimming team, the college newspaper and the football band.

Townes completed work for the Master of Arts degree in Physics at Duke University in 1936, and then entered graduate school at the California Institute of Technology, where he received the Ph.D. degree in 1939 with a thesis on isotope separation and nuclear spins.[1]

A member of the technical staff of Bell Telephone Laboratories from 1933 to 1947, Townes worked extensively during World War II in designing radar bombing systems and has a number of patents in related technology. From this he turned his attention to applying the microwave technique of wartime radar research to spectroscopy, which he foresaw as providing a powerful new tool for the study of the structure of atoms and molecules and as a potential new basis for controlling electromagnetic waves.

At Columbia University, where he was appointed to the faculty in 1948, he continued research in microwave physics, particularly studying the interactions between microwaves and molecules, and using microwave spectra for the study of the structure of molecules, atoms, and nuclei. In 1951, Townes conceived the idea of the maser, and a few months later he and his associates began working on a device using ammonia gas as the active medium. In early 1954, the first amplification and generation of electromagnetic waves by stimulated emission were obtained. Townes and his students coined the word "maser" for this device, which is an acronym for microwave amplification by stimulated emission of radiation. In 1958, Townes and his brother-in-law, Dr. Arthur Leonard Schawlow, for some time a professor at Stanford University but now deceased, showed theoretically that masers could be made to operate in the optical and infrared region and proposed how this could be accomplished in particular systems. This work resulted in their joint paper on optical and infrared masers, or lasers (light amplification by stimulated emission of radiation). Other research has been in the fields of nonlinear optics, radio astronomy, and infrared astronomy. He and his assistants detected the first complex molecules in the interstellar medium and first measured the mass of the black hole in the center of our galaxy.

Having joined the faculty at Columbia University as Associate Professor of Physics in 1948, Townes was appointed Professor in 1950. He served as Executive Director of the Columbia Radiation Laboratory from 1950 to 1952 and was Chairman of the Physics Department from 1952 to 1955.

From 1959 to 1961, he was on leave of absence from Columbia University to serve as Vice President and Director of Research of the Institute for Defense Analyses in Washington, D.C., a nonprofit organization which advised the U.S. government and was operated by eleven universities.

In 1961, Townes was appointed Provost and Professor of Physics at the Massachusetts Institute of Technology (M.I.T). As Provost he shared with the President responsibility for general supervision of the educational and research programs of the Institute. In 1966, he became Institute Professor at M.I.T., and later in the same year resigned from the position of Provost in order to return to more intensive research, particularly in the fields of quantum electronics and astronomy. He was appointed University Professor at the University of California in 1967. In this position Townes is participating in teaching, research, and other activities on several campuses of the University, although he is located at the Berkeley campus.

Dr. Charles Hard Townes statue facing Falls Park on the Reedy River, Greenville, South Carolina

During 1955 and 1956, Townes was a Guggenheim Fellow and a Fulbright Lecturer, first at the University of Paris and then at the University of Tokyo. He was National Lecturer for Sigma Xi and also taught during summer sessions at the University of Michigan and at the Enrico Fermi International School of Physics in Italy, serving as Director for a session in 1963 on coherent light. In the fall of 1963, he was Scott Lecture at the University of Toronto. More recently (2002–2003) he has been the Karl Schwarzschild Lecturer in Germany and the Birla Lecturer and Schroedinger Lecturer in India.

In addition to the Nobel Prize, Townes has received the Templeton Prize, for contributions to the understanding of religion, and a number of other prizes as well as 27 honorary degrees from various universities.

Townes has served on a number of scientific committees advising governmental agencies and has been active in professional societies. This includes being a member, and vice chairman, of the Science Advisory Committee to the President of the U.S., Chairman of the Advisory Committee for the first human landing on the moon, and chairman of the Defense Department's Committee on the MX missile. He also served on the boards of General Motors and of the Perkin Elmer Corporation.

Townes and his wife (the former Frances H. Brown; they married in 1941) live in Berkeley, California. They have four daughters: Linda Rosenwein, Ellen Anderson, Carla Kessler, and Holly Townes.

NIBIB Director Dr. Roderic Pettigrew (left) with Dr. Charles Townes

Research

Charlie Townes was the lead researcher in the construction of the Infrared Spatial Interferometer, the first astronomical interferometer to operate in the mid-infrared. He continues researching into astrophysics and astronomy at the University of California, Berkeley. With Arthur Leonard Schawlow, he wrote the book *Microwave Spectroscopy*, published in 1955.

During his time at Bell Labs Townes was asked to help with the development of a new radar system for aircraft in World War II. He never served in the military, but felt he was helping his country from within the lab. Townes and his team were successful in creating more accurate and precise radar systems, but none of them were ever mass produced by the military. Some of the new systems developed were used as prototypes in early B-52 bombers. After the war, Townes continued to work at Bell Labs, creating new radar by experimenting with different radio wavelengths. Moving from Bell Labs in 1948, to the physics department of Columbia University allowed Townes to return to experimental physics and away from the applications of physics. At Columbia, his research was still partially funded by the US Navy's desire for even smaller radar. At Bell Labs Townes helped develop a radar system with a 1.25 centimeter wavelength. After moving to Columbia, the military wanted radar systems with wavelengths only a few millimeters. The shortening of the wavelength led Townes and his colleagues to focus on microwave research. In 1951, the idea of the maser was proposed to Townes' superiors. After three years and many experiments, Townes and Jim Gordon created a working maser.

Science and religion

A member of the United Church of Christ, Townes considers that "science and religion [are] quite parallel, much more similar than most people think and that in the long run, they must converge".[2] In 2005, he was awarded the Templeton Prize for "Progress Toward Research or Discoveries about Spiritual Realities."

Awards

Townes has been widely recognized for his scientific work and leadership.

- 1956 - elected Full Member of the National Academy of Sciences.
- 1957 - elected Fellow of the American Academy of Arts and Sciences.[3]
- 1958 - awarded the Comstock Prize in Physics from the National Academy of Science.[4]
- 1961 - awarded the David Sarnoff Electronics Award given by the Institute of Electrical and Electronics Engineers, and the Rumford Medal awarded by the American Academy of Arts and Sciences.
- 1962 - The John J. Carty Award for the Advancement of Science given by the National Academy of Science.[5]
- 1962 - The Stuart Ballantine Medal given by The Franklin Institute.

- 1963 - Young Medal and Prize, for distinguished research in the field of optics presented by the Institute of physics.
- 1964 - Nobel Prize in Physics with N. G. Basov and Aleksandr Prokhorov for contributions to fundamental work in quantum electronics leading to the development of the maser and laser.
- 1979 - He was awarded the Niels Bohr international medal awarded for contributions to the peaceful use of atomic energy.
- 1980 - Townes was inducted by his home state into the South Carolina Hall of Science and Technology, and has also been awarded a South Carolina Hall of Science and Technology Citation.
- 1982 - He received the National Medal of Science, presented by President Ronald Reagan.
- 1994 - elected Foreign Member of the Russian Academy of Sciences.
- 1996 - awarded the Frederic Ives Medal [6] by the OSA
- 1998 - awarded the Henry Norris Russell Lectureship by the American Astronomical Society.
- 2000 - awarded the Lomonosov Medal by the Russian Academy of Sciences.
- 2003 - awarded the Telluride Tech Festival Award of Technology in Telluride, Colorado.
- 2005 - awarded the Templeton Prize for "Progress Toward Research or Discoveries about Spiritual Realities."
  - He has also been awarded the LeConte Medallion.
- 2006 - Along with associate Raj Reddy, Townes was awarded the Vannevar Bush Award for Lifetime Contributions and Statesmanship to Science
- 2008 - On May 24 Townes received an Honorary Doctorate of Humane Letters from the University of Redlands.
- 2010 - SPIE Gold Medal

Representation

- Between 1966 and 1970 he was chairman of the NASA Science Advisory Committee for the Apollo lunar landing program.

Bibliography

- M. Bertolotti, *History of the Laser*, Taylor and Francis, 2004.
- J.L. Bromberg, *The Laser in America, 1950-1970*, MIT Press, 1991.
- R.Y. Chiao, *Amazing Light : A Volume Dedicated To Charles Hard Townes On His 80th Birthday*, Springer, 1996.
- J. Hecht, *Beam: The Race to Make the Laser*, Oxford University Press, 2005.
- J. Hecht, *Laser Pioneers*, Academic Press, 1991.
- N. Taylor, *Laser: The Inventor, the Nobel Laureate, and the Thirty-Year Patent War*, Citadel, 2003.
- A.L. Schawlow and C.H. Townes, "Infrared and Optical Masers," Phys. Rev. 112, 1940 (1958).
- C.H. Townes, *Making Waves*, AIP Press, 1995.
- C.H. Townes, *How the Laser Happened: Adventures of a Scientist*, Oxford University Press, 2000.
- C.H. Townes and A.L. Schawlow, *Microwave Spectroscopy*, McGraw-Hill, 1955.
- F. Townes, *Misadventures of a Scientist's Wife*, Regent Press, 2007.

See also

- List of science and religion scholars

References

[1] IEEE Global History Network (1992). "Charles Townes Oral History" (http://www.ieeeghn.org/wiki/index.php/Oral-History:Charles_H.
Townes(1992)). IEEE History Center. . Retrieved 14 July 2011.

[2] Harvard Gazette June 16, 2005 Laser's inventor predicts meeting of science, religion (http://news.harvard.edu/gazette/2005/06.16/
05-laser.html)

[3] "Book of Members, 1780-2010: Chapter T" (http://www.amacad.org/publications/BookofMembers/ChapterT.pdf). American Academy
of Arts and Sciences. . Retrieved 7 April 2011.

[4] "Comstock Prize in Physics" (http://www.nasonline.org/site/PageServer?pagename=AWARDS_comstock). National Academy of
Sciences. . Retrieved 13 February 2011.

[5] "John J. Carty Award for the Advancement of Science" (http://www.nasonline.org/site/PageServer?pagename=AWARDS_carty).
National Academy of Sciences. . Retrieved 13 February 2011.

[6] http://www.osa.org/aboutosa/awards/osaawards/awardsdesc/ivesquinn/

External links

- http://news.harvard.edu/gazette/2005/06.16/05-laser.html
- Charles Hard Townes (http://nobelprize.org/physics/laureates/1964/townes-bio.html)
- Amazing Light: Visions of Discovery (Symposium in honor of Charles Townes) (http://www.metanexus.net/
fqx/townes)
- Bright Idea: The First Lasers (history with interview clips) (http://www.aip.org/history/exhibits/laser/
sections/raydevices.html)
- Infrared Spatial Interferometer Array (http://isi.ssl.berkeley.edu/)
- Research page (http://physics.berkeley.edu/research/faculty/Townes.html)
- Oral History interview transcript with Charles H. Townes 20 and 21 May 1987, American Institute of Physics,
Niels Bohr Library and Archives (http://www.aip.org/history/ohilist/4918.html)
- Dedication Program for the Charles H. Townes Center for Science, Furman University, November 1, 2008 (http:/
/yearofthesciences.furman.edu/Townes_Center_Dedication_program.pdf)

Maser

A **maser** is a device that produces coherent electromagnetic waves through amplification by stimulated emission. Historically, "maser" derives from the original, upper-case acronym **MASER**, which stands for "**M**icrowave **A**mplification by **S**timulated **E**mission of **R**adiation". The lower-case usage arose from technological development having rendered the original denotation imprecise, because contemporary masers emit EM waves (microwave and radio frequencies) across a broader band of the electromagnetic spectrum; thus, the physicist Charles H. Townes's suggested usage of "molecular" replacing "microwave", for contemporary linguistic accuracy.[1] In 1957, when the optical coherent oscillator was first developed, it was denominated *optical maser*, but usually called laser (Light Amplification by Stimulated Emission of Radiation), the acronym Gordon Gould established in 1957.

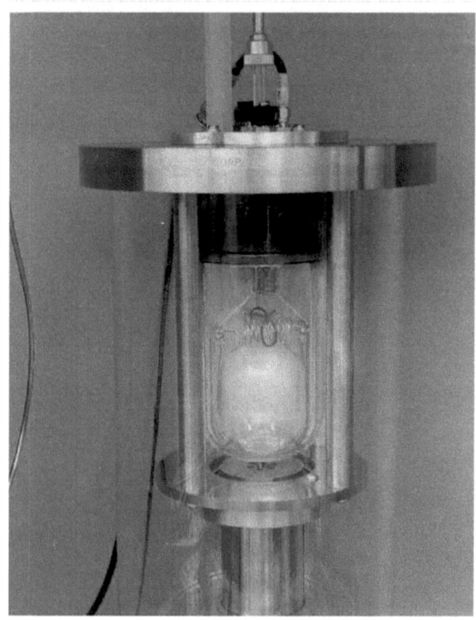

A hydrogen radio frequency discharge, the first element inside a hydrogen maser (see description below)

History

Theoretically, the principle of the maser was described by Nikolay Basov and Alexander Prokhorov from Lebedev Institute of Physics at an *All-Union Conference on Radio-Spectroscopy* held by USSR Academy of Sciences in May 1952.

The results were subsequently published in October 1954. A precursor of the maser was the first *show boosted* hydrogen device built and tested by physicists Theodor V. Ionescu and Vasile Mihu in 1946. Independently, Charles H. Townes, J. P. Gordon, and H. J. Zeiger built the first ammonia maser at Columbia University in 1953. The device used stimulated emission in a stream of energized ammonia molecules to produce amplification of microwaves at a frequency of 24 gigahertz.

Townes later worked with Arthur L. Schawlow to describe the principle of the *optical maser*, or *laser*, which Theodore H. Maiman first demonstrated in 1960. For their research in this field Townes, Basov, and Prokhorov were awarded the Nobel Prize in Physics in 1964.

Technology

The maser is based on the principle of stimulated emission proposed by Albert Einstein in 1917. When atoms have been put into an excited energy state, they can amplify radiation at the proper frequency. By putting such an amplifying medium in a resonant cavity, feedback is created that can produce coherent radiation.

Some common types of masers

- Atomic beam masers
  - Ammonia maser
  - Free Electron Maser
  - Hydrogen maser
- Gas masers
  - Rubidium maser
- Solid State masers
  - Ruby maser

The dual noble gas maser is an example of a masing medium which is nonpolar.[2]

Uses

Masers serve as high precision frequency references. These "atomic frequency standards" are one of the many forms of atomic clocks. They are also used as electronic amplifiers in radio telescopes. Masers are being developed as directed-energy weapons.

Hydrogen maser

Today, the most important type of maser is the hydrogen maser which is currently used as an atomic frequency standard. Together with other types of atomic clocks, they constitute the "Temps Atomique International" or TAI. This is the international time scale, which is coordinated by the Bureau International des Poids et Mesures, or BIPM.

It was Norman Ramsey and his colleagues who first realized this device. Today's masers are identical to the original design. The maser oscillation relies on stimulated emission between two hyperfine levels of atomic hydrogen. Here is a brief description of how it works:

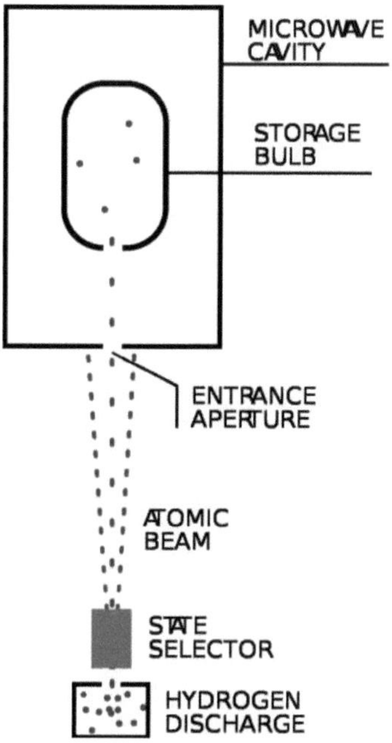

- First, a beam of atomic hydrogen is produced. This is done by submitting the gas at low pressure to an RF discharge (see the picture on this page).

- The next step is "state selection"—in order to get some stimulated emission, it is necessary to create a population inversion of the atoms. This is done in a way that is very similar to the famous Stern-Gerlach experiment. After passing through an aperture and a magnetic field, many of the atoms in the beam are left in the upper energy level of the lasing transition. From this state, the atoms can decay to the lower state and emit some microwave radiation.

A hydrogen maser.

- A high quality factor microwave cavity confines the microwaves and reinjects them repeatedly into the atom beam. The stimulated emission amplifies the microwaves on each pass through the beam. This combination of amplification and feedback is what defines all oscillators. The resonant frequency of the microwave cavity is exactly tuned to the hyperfine structure of hydrogen: 1420 405 751.768 Hz.

- A small fraction of the signal in the microwave cavity is coupled into a coaxial cable and then sent to a coherent receiver.

- The microwave signal coming out of the maser is very weak (a few pW). The frequency of the signal is fixed and *extremely* stable. The coherent receiver is used to amplify the signal and change the frequency. This is done using a series of phase-locked loops and a high performance quartz oscillator.

Astrophysical masers

Maser-like stimulated emission also occurs in nature in interstellar space, and is frequently called superradiant emission to distinguish it from laboratory masers. Such emission is observed from molecules such as water (H_2O), hydroxyl radicals (OH), methanol (CH_3OH), formaldehyde (CH_2O), and silicon monoxide (SiO). Water molecules in star-forming regions can undergo a population inversion and emit radiation at 22 GHz, creating the brightest spectral line in the radio universe. Some water masers also emit radiation from a vibrational mode at 96 GHz.

Extremely powerful masers, associated with active galactic nuclei, are known as megamasers and are up to a million times more powerful than stellar masers.

Terminology

The meaning of the term *maser* has changed slightly since its introduction. Initially the acronym was universally given as "microwave amplification by stimulated emission of radiation," which described devices which emitted in the microwave region of the electromagnetic spectrum.

The principle and concept of stimulated emission has since been extended to more devices and frequencies. Thus the original acronym is sometimes modified, as suggested by Charles H. Townes,[1] to "*molecular* amplification by stimulated emission of radiation." Some have asserted that Townes's efforts to extend the acronym in this way were primarily motivated by the desire to increase the importance of his invention, and his reputation in the scientific community.[3]

When the laser was developed, Townes and Schawlow and their colleagues at Bell Labs pushed the use of the term *optical maser*, but this was largely abandoned in favor of *laser*, coined by their rival Gordon Gould.[4] In modern usage, devices that emit in the X-ray through infrared portions of the spectrum are typically called lasers, and devices that emit in the microwave region and below are commonly called *masers*, regardless of whether they emit microwaves or other frequencies.

Gould originally proposed distinct names for devices that emit in each portion of the spectrum, including *grasers* (gamma ray lasers), *xasers* (x-ray lasers), *uvasers* (ultraviolet lasers), *lasers* (visible lasers), *irasers* (infrared lasers), *masers* (microwave masers), and *rasers* (RF masers). Most of these terms never caught on, however, and all have now become (apart from in science fiction) obsolete except for *maser* and *laser*.

See also

- Masers in science fiction
- Spaser
- List of laser types

References

[1] Charles H. Townes – Nobel Lecture (http://nobelprize.org/physics/laureates/1964/townes-lecture.pdf)

[2] The Dual Noble Gas Maser (http://cfa-www.harvard.edu/Walsworth/Activities/DNGM/old-DNGM.html), Harvard University, Department of Physics

[3] Taylor, Nick (2000). *LASER: The inventor, the Nobel laureate, and the thirty-year patent war*. New York: Simon & Schuster. ISBN 0-684-83515-0.

[4] Taylor, Nick (2000). *LASER: The inventor, the Nobel laureate, and the thirty-year patent war*. New York: Simon & Schuster. pp. 66–70. ISBN 0-684-83515-0.

Further reading

- J.R. Singer, *Masers*, John Whiley and Sons Inc., 1959.
- J. Vanier, C. Audoin, *The Quantum Physics of Atomic Frequency Standards*, Adam Hilger, Bristol, 1989.
- Cartoon Megas XLR 2005

External links

- arXiv.org search for "maser" (http://arxiv.org/find/grp_physics/1/ti:+maser/0/1/0/all/0/1?per_page=100;)
- Noble gas Maser (http://www.cfa.harvard.edu/Walsworth/Activities/DNGM/old-DNGM.html)
- "The Hydrogen Maser Clock Project" (http://web.archive.org/web/20061010151459/http://www.cfa. harvard.edu/hmc/). Harvard-Smithsonian Center for Astrophysics. Archived from the original (http://cfa-www. harvard.edu/hmc/) on 2006-10-10.
- Bright Idea: The First Lasers (http://www.aip.org/history/exhibits/laser)

- Invention of the Maser and Laser (http://focus.aps.org/story/v15/st4), American Physical Society
- Shawlow and Townes Invent the Laser (http://www.bell-labs.com/about/history/laser/), Bell Labs

rue:Macep

Windsor,_Ontario

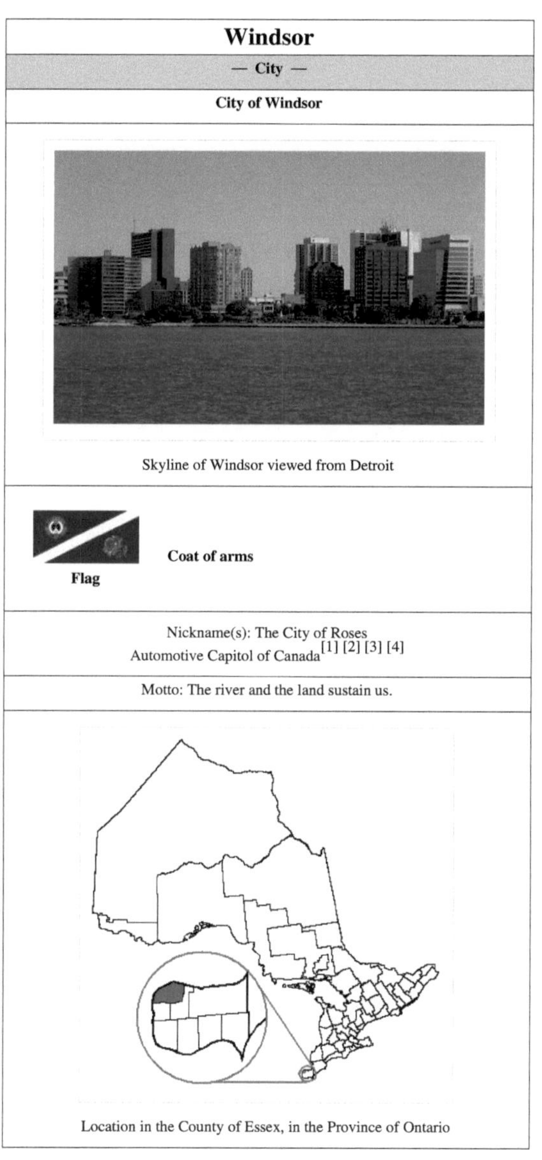

Windsor

— City —

City of Windsor

Skyline of Windsor viewed from Detroit

Flag

Coat of arms

Nickname(s): The City of Roses
Automotive Capitol of Canada[1] [2] [3] [4]

Motto: The river and the land sustain us.

Location in the County of Essex, in the Province of Ontario

Coordinates: 42°17′N 83°00′W	
Country	▌✦▌ Canada
Province	▨▨ Ontario
Census division	Essex *
Settled	1749
Incorporated	1854
Government	
• **Type**	Council-Manager
• **Mayor**	Eddie Francis
• **Governing body**	Windsor City Council
• **CAO**	Helga Reidel
• **MPs**	Joe Comartin (NDP) Brian Masse (NDP)
• **MPPs**	Dwight Duncan (LIB) Teresa Piruzza (LIB)
Area [5] [5]	
• **Land**	146.91 km^2 (56.7 sq mi)
• **Metro**	1022.84 km^2 (394.9 sq mi)
Elevation	190 m (623 ft)
Population (2006) [5] [5]	
• **City**	216473
• **Density**	1473.5/km^2 (3816.3/sq mi)
• **Metro**	323342
• **Metro density**	316.1/km^2 (818.7/sq mi)
	(Ranked 20th most populous in Canada) Source: Stats Canada
Time zone	Eastern (EST) (UTC−5)
• **Summer (DST)**	EDT (UTC−4)
Postal code span	N8P to N8T, N8W to N9G
Area code(s)	519 and 226
Website	www.citywindsor.ca [6]
* Separated municipality of Essex County.	

Windsor is the southernmost city in Canada and is located in Southwestern Ontario at the western end of the heavily populated Quebec City − Windsor Corridor. It is within the Essex, Ontario Census division, although administratively separated from the county government. Separated by the Detroit River, Windsor is located south of Detroit, Michigan in the United States. Windsor is known as The City of Roses and residents are known as Windsorites.

History

Mackenzie Hall

Underground Railroad monument Windsor, Ontario

Prior to European exploration and settlement, the Windsor area was inhabited by the First Nations and Native American people. Windsor was settled by the French in 1749 as an agricultural settlement. It is the oldest continually inhabited European settlement in Canada west of Montreal. The area was first named *Petite Côte* ("Little Coast" - as opposed to the longer coastline on the Detroit side of the river). Later it was called *La Côte de Misère* ("Poverty Coast") because of the sandy soils near LaSalle.

Windsor's French Canadian heritage is reflected in many French street names, such as Ouellette, Pelissier, François, Pierre, Langlois, Marentette, and Lauzon. The current street system of Windsor (a grid with elongated blocks) reflects the Canadien method of agricultural land division, where the farms were long and narrow, fronting along the river. Today, the north-south street name often indicates the name of the family that at one time farmed the land. The street system of outlying areas is consistent with the British system for granting land concessions. There is a significant French-speaking minority in Windsor and the surrounding area, particularly in the Lakeshore, Tecumseh and LaSalle areas.

Duff-Baby House

In 1794, after the American Revolution, the settlement of "Sandwich" was founded. It was later renamed to Windsor, after the town in Berkshire, England. The Sandwich neighbourhood on Windsor's west side is home to some of the oldest buildings in the city, including Mackenzie Hall, originally built as the Essex County Courthouse in 1855. Today, this building functions as a community centre. The oldest building in the city is the Duff-Baby House built in 1792. It is owned by Ontario Heritage Trust and houses government offices. The François Baby House in downtown Windsor was built in 1812 and houses Windsor's Community Museum, dedicated to local history.

The City of Windsor was the site of the Battle of Windsor during the Upper Canada Rebellion in 1837. It was also a part of the Patriot War, later that year.

Windsor was established as a village in 1854 (the same year the village was connected to the rest of Canada by the Grand Trunk Railway/Canadian National Railway), then became a town in 1858, and ultimately gained city status in 1892.

A fire consumed much of Windsor's downtown core on October 12, 1871, destroying over 100 buildings.[7]

On October 25, 1960, a massive gas explosion destroyed the building housing the Metropolitan Store on Ouellette Avenue. Ten people were killed and at least one hundred injured.[8] The 45th anniversary of the event was commemorated by the *Windsor Star* on October 25, 2005. It was featured on History Television's Disasters of the Century.

The *Windsor Star* Centennial Edition in 1992 covered the city's past, its success as a railway centre, and its contributions to World War I and World War II. It also recalled the naming controversy in 1892 when the town of Windsor aimed to become a city. The most popular names listed in the naming controversy were "South Detroit", "The Ferry" (from the ferries that linked Windsor to Detroit), Windsor, and Richmond (the runner-up in popularity). Windsor was chosen to promote the heritage of new English settlers in the city and to recognize Windsor Castle in Berkshire, England. However, Richmond was a popular name used until the Second World War, mainly by the local post office.

Sandwich, Ford City and Walkerville were separate legal entities (towns) in their own right until 1935. They are now historic neighbourhoods of Windsor. Ford City was officially incorporated as a village in 1912; it became a town in 1915, and a city in 1929. Walkerville was incorporated as a town in 1890. Sandwich was established in 1817 as a town with no municipal status. It was incorporated as a town in 1858 (the same year as neighbouring Windsor).

These three towns were each annexed by Windsor in 1935. The nearby villages of Ojibway and Riverside were incorporated in 1913 and 1921 respectively. Both were annexed by Windsor in 1966.[9]

Climate

Windsor has a humid continental climate (Köppen climate classification *Dfa*) with four distinct seasons. The mean annual temperature is 9.5°C (49°F), among the warmest in Canada primarily due to its hot summers. Some locations in coastal and lower mainland British Columbia have a slightly higher mean annual temperature due to milder winter conditions. The coldest month is January and the warmest month is July. The coldest temperature ever recorded in Windsor was −29.1 °C (−20.4 °F) and the warmest was 40.2 °C (104.4 °F).[10]

Summers are hot and humid, and the annual average rain is 94 cm (37 inches). Winters are generally cold with occasional mild periods. Windsor is not located in the lake effect snowbelts and snow cover is intermittent throughout the winter lasting 52 days of snow on the ground compared to 88 days of snow in the ground in London Ontario.; nevertheless, there are typically two to five major snowfall events each winter. Summers are warm/hot and humid with humidex reaching 30 Celsius or above 69 times in a average summer, and thunderstorms are common every 5 days or so. Windsor has the highest number of days per year with lightning, haze, and daily maximum temperatures over 30 °C (86 °F) of cities in Canada. Windsor is also home to Canada's warmest fall, with highest average temperatures for the months of September, October and November. Precipitation is generally well-distributed throughout the year. There are on average 2,265 sunshine hours per year in Windsor.[11]

Tornadoes

The strongest and deadliest tornado to touch down in Windsor was a category F4 in 1946. Windsor was the only Canadian city to experience a tornado during the Super Outbreak of 1974, an F3 which killed nine people at the Windsor Curling Club. The city was grazed by the 1997 Southeast Michigan tornado outbreak with one tornado (an F1) forming east of the city. Tornadoes have been recorded crossing the Detroit River (in 1946 and 1997), and waterspouts are regularly seen over Lake St. Clair and Lake Erie especially in autumn.

On April 25, 2009, an F0 tornado briefly touched down in the city's east end causing minor damage to nearby buildings, most notably a CUPE union hall.[12]

Air pollution

Respiratory illnesses that are associated with pollution are more prevalent here than elsewhere in Canada as Windsor is downwind from several strong polluters.[14]

Windsor Air Quality Study 2010 to 2011

The Weather Network has designated Windsor as "the smog capital of Canada."[15] Windsor's Citizens Environment Alliance holds a yearly art event entitled Smogfest to raise awareness of air quality issues.

A 2001 article in *Environmental Health Perspectives* stated that the rates of mortality, morbidity as hospitalizations, and congenital anomalies in the Windsor Area of Concern ranked among the highest of the 17 Areas of Concern on the Canadian side of the Great Lakes for selected end points that might be related to pollution.[16]

In the summer of 2003, Transit Windsor provided free transit on smog advisory days. The pilot project was extremely successful and drew interest from across the country and Europe. Ridership increased nearly 50% on those days. There was extensive local media coverage, stories on the project were featured on The Weather Network, CBC NewsWorld, in newspapers and on radio stations across the nation.[17] Despite the success, the pilot project was discontinued, as the budget for the program was quickly expended.

Cityscape

Windsor's Department of Parks and Recreation[18] maintains 3000 acres (12 km^2) of green space, 180 parks, 40 miles (64 km) of trails, 22 miles (35 km) of sidewalk, 60 parking lots, vacant lands, natural areas and forest cover within the city of Windsor. The largest park is Mic Mac Park, which can accommodate many different activities including baseball, soccer, biking, and sledding. Windsor has numerous bike trails, the largest being the Ganatchio Trail on the far east side of the city. In recent years, city council has pushed for the addition of bicycle lanes on city streets to provide links throughout the existing trail network.

Windsor's Riverside Drive and Riverfront Bike Trail from Dieppe Gardens.

The Windsor trail network is linked to the LaSalle Trail in the west end, and will eventually be linked to the Chrysler Canada Greenway (part of the Trans Canada Trail). The current greenway is a 42 km former railway corridor that has been converted into a multi-use recreational trail, underground utility corridor and natural green space. The corridor begins south of Oldcastle and continues south through McGregor, Harrow, Kingsville, and Ruthven. The Greenway is a fine trail for hiking, biking, running, birding, cross country skiing and in some areas, horseback riding. It connects natural areas, rich agricultural lands, historically and architecturally significant structures, and award winning wineries. A separate 5 km landscaped traverses the riverfront between downtown and the Ambassador Bridge. Part of this trail winds through Odette Sculpture Park, displaying various modern and post-modern sculptures from artists in Essex County. Families of elephants (see picture), penguins, horses, and many other themed sculptures are found in the park.

Economy

Windsor's economy is primarily based on education, manufacturing, tourism, and government services.

Both the University of Windsor and St. Clair College are significant local employers and have enjoyed substantial growth and expansion in recent years. The recent addition of a full-program satellite medical school of the University of Western Ontario, which opened in 2008 at the University of Windsor is further enhancing the region's economy and the status of the university. The university is currently constructing a $112 million facility for their Engineering Faculty.

Windsor has a well-established tourism industry. Caesars Windsor (formerly Casino Windsor), one of the largest casinos in Canada, ranks as one of the largest local employers. It has been a major draw for U.S. visitors since opening in 1994. Further, the 1150-kilometre (710 mi) Quebec City – Windsor Corridor contains 18 million people, with 51% of the Canadian population and three out of the four largest metropolitan areas, according to the 2001 Census.

The city also boasts an extensive riverfront parks system and fine restaurants, such as those on Erie Street in Windsor's Little Italy called "Via Italia", another popular tourist destination. The Lake Erie North Shore Wine Region in Essex County has enhanced tourism in the region.

Windsor is the headquarters of Hiram Walker & Sons Limited, now owned by Pernod Ricard. Its historic distillery was founded by Hiram Walker in 1858 in what was then Walkerville, Ontario.

Windsor is one of Canada's major automobile manufacturing centres. However, plant closures and significant job losses in recent years have impacted Windsor's automotive manufacturing industry. The city is home to the headquarters of Chrysler Canada. Automotive facilities include the Chrysler minivan assembly plant, two Ford Motor Company engine plants, and a number of tool and die and automotive parts manufacturers.

The city's diversifying economy is also represented by companies involved in pharmaceuticals, alternative energy, insurance, internet and software. Windsor is also home to the Windsor Salt Mine and the Great Lakes Regional office of the International Joint Commission.

Windsor was recently listed as the number two large city for economic potential in North-America and number 7 large city of the future in North America according to the *FDI North-American cities of the future list*. (*American Cities of the Future 2011/12*)

Demographics

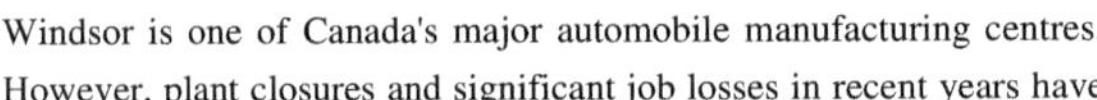

Census	Population
1841	300
1871	4,253
1881	6,561
1891	10,322
1901	12,153
1911	17,829
1921	38,591
1931	63,108

1941	104,415
1951	120,049
1961	114,367
1971	203,300
1981	192,083
1991	191,435
2001	208,402
2006	216,473

Ethnic Origin, 2001[19]	
Ethnic Origin	Percentage
Canadian	28.1%
French	21.2%
English	18.5%
Irish	13.1%
Scottish	12.1%
Italian	9.7%
German	7.1%
Polish	4.0%
Lebanese	2.9%
Ukrainian	2.9%
multiple responses included	

Religion, 2001[20]	
Religion	Percentage
Catholic	48.3%
Protestant	23.9%
No religion	12.1%
Muslim	4.8%
Orthodox	4.3%

In 2006, the population of Windsor was 216,473 and that of the Windsor metropolitan area (consisting of Windsor, Tecumseh, Amherstburg, LaSalle and Lakeshore) was 323,342.[21] This represents a growth of 3.5% in the city population since 2001 and a growth of 5.0% in the metropolitan area population since 2001.[22] During the same period, Ontario grew by 6.1% and Canada by 5.4%.[23]

Because of its jobs, Windsor attracts many immigrants from around the world. Over 20% of the population is foreign-born; this is the fourth-highest proportion for a Canadian city. Visible minorities make up 21.0% of the population, making it the most diverse city in Ontario outside of the Greater Toronto Area.[24] [25]

From the 2001 Canadian census, Windsor's population was 48.9% male and 51.1% female. Children under five accounted for 6.3% of the city population compared to 5.6% for Canada. Persons of retirement age (65 years and over) accounted for 14.1% of the population in Windsor compared to 13.0% for Canada. The median age in Windsor is 36.0 years compared to 37.6 years for Canada.[26]

Government

Windsor City Hall.

Windsor's history as an industrial centre has given the New Democrats (a party partially founded, governed and supported by labour unions), a dedicated voting base. During federal and provincial elections, Windsorites have maintained its local representation in the respective legislatures. The Liberal Party of Canada also has a strong electoral history in the city. Canada's 21st Prime Minister Paul Martin was born in Windsor. His father Paul Martin (Sr.), a federal cabinet minister in several portfolios through the Liberal governments of the 1940s, 1950s and 1960s, was first elected to the House of Commons from a Windsor riding in the 1930s. Martin (Sr.) practiced law in the city and the federal building on Ouellette Avenue is named after him. Eugene Whelan was a Liberal cabinet minister and one-time Liberal party leadership candidate elected from Essex County from the 1960s to the early 1980s, as well as Mark MacGuigan of Windsor-Walkerville riding, who also served as External Affairs, and later Justice minister in the early 1980s. Deputy Prime Minister Herb Gray represented Windsor as an MP from 1962 through 2003, winning thirteen consecutive elections making him the longest serving MP in Canadian history.[27] A bust of Herb Gray is located at the foot of Ouellette Avenue near Dieppe Park in downtown Windsor.

Current representation

The current mayor of Windsor is Eddie Francis, a Lebanese Canadian. He was the city's youngest mayor when he was first elected at age 29 in 2003. Windsor is governed under the Council-Manager form of local government and includes the elected City Council, mayor, and an appointed Chief Administrative Officer. The city is divided into ten wards, with one councillor representing each ward. The mayor serves as the chief executive officer of the city and functions as its ceremonial head. Day-to-day operations of the government are carried out by the Chief Administrative Officer. In August 2009, Windsor City Council approved a 10-ward electoral system for the 2010 civic election. Under the new plan, voters will elect one Councillor in each of the ten new wards. The new election map will double the number of wards that have existed along unchanged boundaries for 30 years.[28]

At the provincial and federal levels, Windsor is divided into two ridings: Windsor West and Windsor—Tecumseh. The city is currently represented in the Legislative Assembly of Ontario by two Liberal MPPs: Teresa Piruzza (Windsor West), and Dwight Duncan (Windsor—Tecumseh).

Federally, Windsor West was a longtime Liberal stronghold under Herb Gray, while Windsor—Tecumseh has traditionally been a Liberal-NDP swing riding. Both ridings are currently represented in the federal Parliament by NDP MPs: Brian Masse (Windsor West) and Joe Comartin (Windsor—Tecumseh).

New city wards

Ward 1 : South(west) Windsor including Roseland, Ojibway Park and Windsor Raceway

Ward 2 : West Windsor including University District and Old Sandwich Towne

Ward 3 : Central Windsor including City Centre, Via Italia, and South Central

Ward 4 : Walkerville and South Walkerville

Ward 5 : East Windsor and Ford City including western Fountainbleau

Ward 6 : Old Riverside including Little River Acres (The Villages) and Pilette Village

Ward 7 : Forest Glade and East Riverside

Ward 8 : eastern Fountainbleau including Roseville Gardens; East Windsor

Ward 9 : South Windsor including Devonshire Heights, Windsor Airport and Old Sandwich South

Ward 10: South Cameron and Remington Park communities

Culture and tourism

Caesars Windsor hotel

Art Gallery of Windsor overlooking riverfront rock gardens

Windsor tourist attractions include Caesars Windsor, a lively downtown club scene, Little Italy, the Art Gallery of Windsor, the Odette Sculpture Park, and Ojibway Park. As a border settlement, Windsor was a site of conflict during the War of 1812, a major entry point into Canada for refugees from slavery via the Underground Railroad and a major source of liquor during American Prohibition. Two sites in Windsor have been designated as National Historic Sites of Canada: the Sandwich First Baptist Church, a church established by Underground Railroad refugees, and François Bâby House, an important War of 1812 site now serving as Windsor's Community Museum.[29] [30]

The Capitol Theatre in downtown Windsor had been a venue for feature films, plays and other attractions since 1929, until it declared bankruptcy in 2007. As of 2009 the Capitol Theatre was open, showcasing various features.

Windsor's nickname is the "Rose City" or the "City of Roses" and Windsor has designated a rose known as Liebeszauber (Love's Magic) as the City of Windsor Rose.[31] Windsor is noted for the several large parks and gardens found on its waterfront. The Queen Elizabeth II Sunken Garden is located at Jackson Park in the central part of the city. A World War II era Avro Lancaster was displayed on a stand in the middle of Jackson Park for over four decades but has since been removed for restoration. This park is now home to a mounted Spitfire replica and a Hurricane replica.

Chrysler's Canada HQ in downtown Windsor, as seen from Dieppe Gardens along the riverfront.

Of the parks lining Windsor's waterfront, the largest is the 5 km (three mile) stretch overlooking the Detroit skyline. It extends from the Ambassador Bridge to the Hiram Walker Distillery. The western portion of the park contains the Odette Sculpture Park which features over 30 large-scale contemporary sculptures for public viewing, along with the Canadian Vietnam Veterans Memorial. The central portion contains Dieppe Gardens, Civic Terrace and Festival Plaza, and the eastern portion is home to the Bert Weeks Memorial Gardens. Further east along the waterfront is Coventry Gardens, across from Detroit's Belle Isle. The focal point of this park is the Charles Brooks Memorial Peace Fountain which floats in the Detroit River and has a coloured light display at night. The fountain is the largest of its kind in North America and symbolizes the peaceful relationship between Canada and the United States.

Fireworks at the Windsor-Detroit International Freedom Festival.

Each summer, Windsor co-hosts the two-week-long Windsor-Detroit International Freedom Festival, which culminates in a gigantic fireworks display that celebrates Canada Day and US Independence Day. The fireworks display is among the world's largest and is held on the final Wednesday in June over the Detroit River between the two downtowns. Each year, the event attracts over a million spectators to both sides of the riverfront.

Following the 2008 Red Bull Air Race World Championship in Detroit, Michigan, Windsor successfully put in a bid to become the first Canadian city to host the event. Red Bull touted the 2009 race in Windsor as one of the most exciting in the seven-year history of the Red Bull Air Race World Championship,[32] and on January 22, 2010, it was announced that Windsor will be a host city for the 2010 and 2011 circuits,[33] along with a select group of major international cities that includes Abu Dhabi, United Arab Emirates, Perth, Australia and New York, New York. The event attracted 200,000 fans to the Detroit River waterfront in 2009. The Red Bull air races were cancelled worldwide for 2011.[32]

Windsor has often been the place where many metro Detroiters find what is forbidden in the United States. With a minimum legal drinking age of 21 in Michigan and 19 in Ontario, a number of 19 and 20-year-old Americans frequent Windsor's bars. The city also became a gaming attraction with Caesars Windsor's opening in 1994, five years before casinos opened in Detroit. In addition, one can purchase Cuban cigars, Cuban Rum, less-costly prescription drugs, Absinthe, certain imported foods, and other items not available in the United States. In addition, many same-sex couples from the United States have chosen to marry in Windsor, which is illegal in Michigan and most parts of the United States, but is legal in Canada.

Media

Windsor is considered part of the Detroit television and radio market for purposes of territorial rights. Due to this fact, and its proximity to Toledo and Cleveland, radio and television broadcasters in Windsor are accorded a special status by the Canadian Radio-television and Telecommunications Commission, exempting them from many of the Canadian content ("CanCon") requirements most broadcasters in Canada are required to follow. The CanCon requirements are sometimes blamed in part for the decline in popularity of Windsor radio station CKLW, a 50,000 watt AM radio station that in the late 1960s (prior to the advent of CanCon) had been the top-rated radio station not only in Detroit and Windsor, but also in Toledo and Cleveland.

Windsor Star offices on Ferry Street

Windsor has also been exempt from concentration of media ownership rules. Except for Blackburn Radio-owned stations CJWF-FM and a rebroadcaster of Chatham's CKUE-FM in Windsor, all other current commercial media outlets are owned by a single company, CTVglobemedia.

The city is also home to one campus radio station, CJAM-FM, situated on the University of Windsor campus.[34]

Education

Dillon Hall, University of Windsor

Windsor is home to the University of Windsor, which is Canada's southernmost university. It is a research oriented, comprehensive university with a student population of 16,000 full-time graduate and undergraduate students. Now entering its most ambitious capital expansion since its founding in 1963, the University of Windsor recently opened the Anthony P. Toldo Health Education & Learning Centre, which houses the Schulich School of Medicine & Dentistry. As well, with the help of $40 million in Ontario government funding, the University is currently building a 300000-square-foot (28000 m^2), $112-million Centre for Engineering Innovation; a structure that will establish revolutionary design standards across Canada and beyond. Construction is scheduled to be completed by Fall 2012.

St. Clair College campus on Riverside Drive.

The university is just east of the Ambassador Bridge, south of the Detroit River. Windsor is also home to St. Clair College with a student population of 6500 full-time students. Its main campus is in Windsor, and it also has campuses in Chatham and Wallaceburg. In 2007, St. Clair College opened a satellite campus in downtown Windsor in the former Cleary International Centre. In April 2010, St. Clair College added to its downtown Windsor presence with the addition of its MediaPlex school. Together, they bring over one thousand students into the downtown core every day. More recently Collège Boréal opened an access centre and small campus to their Ouellette avenue location. This small campus offers access to many Collège Boréal programmes as well as immigration and integration assistance for francophones in the area. Collège Boréal is Windsor's only francophone post-secondary institution, providing service for a small, but notable, population of Franco-Ontarien within the Windsor-Tecumseh-Belle River area.

In Spring 2011, it was announced that the University of Windsor would move its music and visual art programs downtown to be housed in the historic Armouries building at Freedom Way and University Ave E. The move should bring an additional 500 students into the downtown core daily. The University is also studying the feasibility of moving the School of Social Work to the Windsor Star buildings at the corner of Pitt and Ferry streets.

Windsor is home to two International Baccalaureate recognized schools: Assumption College School (a Catholic high school) and Académie Ste. Cécile International School (a private school). Vincent Massey Secondary School is renowned in Southern Ontario for its notable accomplishments nationally in mathematics and computer science.

Windsor youth attend schools in the Greater Essex County District School Board, the Windsor-Essex Catholic District School Board, Conseil scolaire de district des écoles catholiques du Sud-Ouest and Conseil scolaire de district du Centre-Sud-Ouest more recently known as Conseil scolaire Viamonde. Independent faith-based schools include Maranatha Christian Academy (JK-12), First Lutheran Christian Academy (preschool-8), and Académie Ste. Cécile International School (JK-12, including International Baccalaureate), and Windsor Adventist Elementary School. The non-denominational Lakeview Montessori School is a private school as well.

The Windsor Public Library offers education, entertainment and community history materials, programs and services. The main branch coordinates a literacy program for adults needing functional literacy upgrading.

The Canada South Science City[35] serves the Elementary School Curriculum's Science and Technology component.

Infrastructure

Health systems

There are two hospitals in Windsor: Hôtel-Dieu Grace Hospital and Windsor Regional Hospital. Hôtel-Dieu Grace Hospital is the result of an amalgamation of Grace Hospital and Hôtel-Dieu in 1994. The merger occurred due to the Government of Ontario's province-wide policy to consolidate resources into Local Health Integrated Networks, or LHINs. This was to eliminate duplicate services and allocate resources more efficiently across the region. The policy resulted in the closure of many community-based and historically important hospitals across the province. Two of Windsor's independent hospitals: Metropolitan General Hospital on Lens Ave and Windsor Western Hospital on Prince Road, were joined to form Windsor Regional Hospital. The original hospital sites remain but the operations are administratively centralized through the new collective structure.

Windsor hospitals have formal and informal agreements with Detroit-area hospitals. For instance, pediatric neurosurgery is no longer performed in Windsor. *The Windsor Star* reported in July 2007 that Hôtel-Dieu Grace has formally instituted an agreement with Detroit's Harper Hospital to provide this specialty and surgery for the dozen patients requiring care annually. Leamington District Memorial Hospital in Leamington, Ontario serves much of Essex County and, along with the Windsor institutions, share resources with the Chatham-Kent Health Alliance.

Over five thousand Windsor residents are employed in the health care industry alone in Metro Detroit. With more work hours and a generally higher rate of pay, there is frustration among Windsor hospital administration to attract and retain skilled nurses and doctors to work in Ontario.

The Essex County Medical Society lists family doctors accepting patients.[36] Many people who do not seek a family doctor use the region's many walk-in clinics for regular medical conditions.

Transportation

See also: Roads in Windsor, Ontario, and Bike trails in Windsor, Ontario.

Highway 401 in Windsor near its western terminus with Dougall Parkway.

Windsor is the western terminus of both Highway 401, Canada's busiest highway, and Via Rail's Quebec City-Windsor Corridor. Windsor's Via station is the nation's sixth-busiest in terms of passenger volumes. The city is served by Windsor Airport with regular, scheduled commuter air service by Air Canada Jazz and heavy general aviation traffic. The Detroit Metropolitan Wayne County Airport is located approximately 40 km across the border in Romulus, Michigan and is the airport of choice for many Windsor residents as it has regular flights to a larger variety of destinations than Windsor Airport.[37] Windsor is also located on the St. Lawrence Seaway, and is accessible to ocean-going vessels.

New bus terminal opened in 2007.

Public Transportation is provided by Transit Windsor, the city-owned bus company, operating several bus routes through the city as well as providing transportation for many of the city's secondary school students. Transit Windsor shares its newly constructed $8-million downtown depot with Greyhound Lines. The new depot opened in 2007. A street car line was proposed in 2011 as a way to increase east to west connections into downtown, but was quickly shot down by Transit Windsor weeks later during a strategic planning session. "Street cars are just too expensive to include in Windsor's transit future."

Windsor has a municipal highway, E.C. Row Expressway, running east-west through the city. Consisting of 15.7 km (9.8 mi) of highway and nine interchanges, the expressway is the fastest way for commuters to travel across the city. E.C. Row Expressway is mentioned in the Guinness Book of Records as the shortest freeway that took the longest time to build as it took more than 15 years to complete. The expressway stretches from Windsor's far west end at Ojibway Parkway east to Banwell Road on the city's border with Tecumseh.

Via rail train at Windsor train station

The majority of development in Windsor stretches along the water instead of in-land. As a result, there is a lack of east-west arteries compared to north-south arteries. Only Riverside Drive, Wyandotte Street, Tecumseh Road and the E.C. Row Expressway serve the almost 30 kilometres (19 mi) from the west end of Windsor eastward. All of these roads are burdened with east-west commuter traffic from the development in the city's east end and suburbs further east. There are eight north-south roads interchanging with the expressway: Huron Church Road, Dominion Boulevard, Dougall Avenue, Howard Avenue, Walker Road, Central Avenue, Jefferson Boulevard, and Lauzon Parkway.

Traffic backups on some of these north-south roads at the E.C. Row Expressway are common, mainly at Dominion, Dougall, Howard, and Walker as the land south of the expressway and east of Walker is occupied by Windsor airport and there has been little development.

Windsor's many rail crossings intersect with these north-south thoroughfares. In October 2008, the Province of Ontario completed a grade separation at Walker Road and the CP Rail line. Another grade separation was completed in November 2010 at Howard Avenue and the CP Rail line. In both cases, the road travels under the rail line and both have below grade intersections with an east-west street. These were planned as parts of the "Let's Get

Windsor-Essex Moving" project funded by the Province of Ontario to improve local transportation infrastructure.

Windsor is connected to Essex and Leamington via Highway 3, and is well connected to the other municipalities and communities throughout Essex County via the county road network. Nearly 20,000 vehicles travel on Highway 3 in Essex County on a daily basis. It is the main route to work for many residents of Leamington, Kingsville and Essex.

Windsor is linked to the United States by the Ambassador Bridge, the Detroit-Windsor Tunnel, a Canadian Pacific Railway tunnel, and the Detroit-Windsor Truck Ferry. The Ambassador Bridge is North America's #1 international border crossing in terms of goods volume: 27% of all trade between Canada and the United States crosses at the Ambassador Bridge.

Windsor has a bike trail network including the (Riverfront Bike Trail, Ganatchio Bike Trail, and Little River Extension). They have become a blend of parkland and transportation, as people use the trails to commute to work or across downtown on their bicycles.

The Port of Windsor is located on the Great Lakes/St. Lawrence Seaway System, on the Detroit River. The port is the third largest Canadian Great Lakes port in terms of shipments.[38]

Ambassador Bridge and potential third crossing

The Ambassador Bridge at sunset.

A major and controversial issue is the amount of traffic to and from the Ambassador Bridge. The number of vehicles crossing the bridge has doubled since 1990. However, the total volume of traffic has been declining since the September 11, 2001 attacks.

Access to the Ambassador Bridge is via two municipal roads: Huron Church Road and Wyandotte Street. A large portion of the traffic consists of tractor-trailers. There have been at times a wall of trucks up to 8 km (5.0 mi) long on Huron Church Road. This road cuts through the west end of the city and the trucks are the source of many complaints about noise, pollution and pedestrian hazards. In 2003, a single mother of three, Jacqueline Bouchard, was struck and killed by a truck at the corner of Huron Church and Girardot Avenue in front of Assumption College Catholic High School, a tragedy argued to be due to a lack of practical safety precautions.[39]

Windsor City Council hired famous traffic consultant Sam Schwartz to produce a proposal for a solution to this traffic problem. City councillors overwhelmingly endorsed the proposal and it was presented to the federal government as a "Made in Windsor" solution. Not all of the surrounding residents supported the plan. One problem with the plan is that the proposed road would cut through protected green spaces such as the Ojibway Prairie Reserve.

In 2005, the Detroit River International Crossing (DRIC - a joint Canadian-American committee studying the options for expanding the border crossing) announced that its preferred option was to extend Highway 401 directly westward to a new bridge spanning the Detroit River and interchange with Interstate 75 somewhere between the existing Ambassador Bridge span and Wyandotte.

On April 9, 2010, the City of Windsor, along with local cabinet ministers Dwight Duncan and Sandra Pupatello of the Province of Ontario, announced that a final decision had been made in the plans to construct the Windsor-Essex Parkway, the new Highway 401 extension leading to a future crossing. The announcement indicated that the project will be the most expensive road ever built in Canada on a per kilometre basis, and included commitments to enhance green space design through the use of berming, landscaping, and other aesthetic treatments. As part of negotiations with the City of Windsor (who threatened legal action in pursuit of more tunneling and green space of the route), the province agreed to additional funding to infrastructure projects in Windsor-Essex; this includes money for the improvement to the plaza of the Canadian side of the Windsor-Detroit tunnel, the widening and other improvements of Walker Rd between Division Rd and E.C. Row Expressway, and the environmental assessment and preliminary design of a future extension of Lauzon Parkway to Highway 401.

Skyline

Sister cities

Windsor has several sister cities in the world - dates are in parentheses:

- Changchun, China (1992)[40]
- Coventry, U.K. (1963)[41]
- Fujisawa, Japan (1987)
- Granby, Quebec, Canada (1956)[42]
- Cornwall, Ontario, Canada, (1972)
- Gunsan, South Korea (2005)[43]
- Lublin, Poland (2000)[44]
- Mannheim, Germany (1980)[45]
- Las Vueltas, El Salvador (1987)[46]
- Ohrid, Macedonia
- Saint-Étienne, France (1963)[41]
- Saltillo, Mexico
- Udine, Italy (1975)[47]

Sports

Windsor's sports fans tend to support the major professional sports league teams in either Detroit or Toronto, but the city itself is home to the following youth, minor league, post-secondary and professional teams. Many Windsor sports teams at the amateur level are sponsored by the AKO Fraternity.

The WFCU Centre is the current home of the Windsor Spitfires.

- *Windsor Spitfires* (Ontario Hockey League Major Junior "A" 2009 & 2010 Memorial Cup Champions)
- *Windsor Clippers* (Ontario Lacrosse Association Junior "B")
- *Windsor AKO Fratmen* (Canadian Junior Football League)
- *Windsor Lancers* (Canadian Interuniversity Sport)
- *St. Clair Saints* [48] (Canadian Colleges Athletic Association)
- *Windsor Rogues Rugby*[49] *(Ontario Rugby Union (ORU))*
- *Windsor FC Nationals* [50] (Ontario Youth Soccer League)(Western Ontario Youth Soccer League)
- *Windsor Fight Team* [51] (Mixed Martial Arts)

Former teams

- *Windsor Bulldogs* (OHA Senior A Hockey League) 1953-1964, won 1963 Allan Cup)
- *Windsor St. Clair Saints* (Major League Hockey Senior "AAA"/CCAA)
- *Windsor Royals/Bulldogs* (Western Ontario Hockey League) now known as LaSalle Vipers
- *Windsor Bulldogs* (Canadian Professional Hockey League) 1920s and 1930s
- *Windsor Hornets* (Canadian Professional Hockey League) 1920s
- *Windsor Border Stars* (Canadian Soccer League)
- *Windsor Gotfredsons* (International Hockey League) 1940s
- *Windsor Spitfires* (International Hockey League) 1940s
- *Windsor Warlocks* (Major Series Lacrosse) 2004
- *Windsor Clippers* (OLA Senior B Lacrosse League) 1960s
- *Windsor Warlocks* (OLA Junior A Lacrosse League) 1970s
- *Windsor Warlocks* (OLA Junior B Lacrosse League) 1980s
- *Windsor AKO Fratmen* (OLA Junior B Lacrosse League) 2003-2009
- *Windsor Mariners* (Ontario Australian Football League) 2000s

Red Bull Air Races

Windsor has hosted a round of the Red Bull Air Race World Championship in each of 2009 and 2010 (Detroit hosted the race in 2008). The races take place on a course of pylons set up on the Detroit River, right over the border between Canada and the USA.

See also

- 1946 Windsor–Tecumseh, Ontario tornado
- Detroit–Windsor
- Flag of Windsor, Ontario
- Super Outbreak

References

[1] Bubbers, Matt. "Canada car capitol named top future city" (http://autos.sympatico.ca/weird-automotive-news/8689/canada-car-capitol-named-top-future-city). . Retrieved 22 December 2011.

[2] "Business And Industry" (http://www.windsorzip.ca/Business_And_Industry.asp). Windsorzip.ca. . Retrieved 2012-01-02.

[3] "In the Beginning" (http://www.windsorpubliclibrary.com/digi/wow/intro.htm). Windsorpubliclibrary.com. . Retrieved 2012-01-02.

[4] Grand Bend Motorplex. "Sponsorship" (http://windsorweekend.ca/sponsorship.html). Windsorweekend.ca. . Retrieved 2012-01-02.

[5] "Windsor (city) community profile" (http://www12.statcan.gc.ca/census-recensement/2006/dp-pd/prof/92-591/details/page.cfm?Lang=E&Geo1=CSD&Code1=3537039&Geo2=PR&Code2=35&Data=Count&SearchText=windsor&SearchType=Begins&SearchPR=01&B1=All&Custom=). *2006 Census data*. Statistics Canada. . Retrieved 2011-04-04.

[6] http://www.citywindsor.ca/

[7] "The Timeline: Fire of 1871" (http://209.202.75.197/digi/chi/timeline.asp?Lang=english). *Settling Canada's South: How Windsor Was Made*. Windsor Public Library. 2002. . Retrieved 2008-03-14.

[8] "History" (http://www.windsorfire.com/ecom.asp?pg=history&specific=17). Windsorfire.com. . Retrieved 2012-01-02.

[9] "City of Windsor: Heritage" (http://www.citywindsor.ca/001952.asp). *citywindsor.ca*. City of Windsor. . Retrieved July 20, 2011.

[10] Environment Canada. Retrieved March 28, 2008. (http://www.climate.weatheroffice.ec.gc.ca/climate_normals/results_e.html?StnID=4716&autofwd=1)

[11] "The Climate and Weather of Windsor, Ontario" (http://www.livingin-canada.com/climate-windsor.html). Livingin-canada.com. 2006-12-03. . Retrieved 2012-01-02.

[12] "Enivronment Canada. Retrieved April 28, 2009" (http://www.weatheroffice.gc.ca/warnings/SWS_bulletins_e.html?prov=on). Weatheroffice.gc.ca. 2011-12-06. . Retrieved 2012-01-02.

[13] Environment Canada— Canadian Climate Normals 1971–2000 (http://www.climate.weatheroffice.ec.gc.ca/climate_normals/results_e.html?Province=ONT&StationName=&SearchType=&LocateBy=Province&Proximity=25&ProximityFrom=City&StationNumber=&IDType=MSC&CityName=&ParkName=&LatitudeDegrees=&LatitudeMinutes=&LongitudeDegrees=&LongitudeMinutes=&

NormalsClass=A&SelNormals=&StnId=4716&). Retrieved 13 August 2009.

[14] Windsor, The (2008-04-27). "Windsor 'the most polluted city in North America': RFK Jr" (http://www.canada.com/windsorstar/story. html?id=4cb4ab4f-772e-47fc-8a01-3e0621454330&k=93563). Canada.com. . Retrieved 2012-01-02.

[15] "Air Quality - Air Quality - A Provincial Prospective" (http://www.theweathernetwork.com/index.php?product=airquality& airqualitycode=on&pagecontent=airqsummary). The Weather Network. . Retrieved 2012-01-02.

[16] Gilbertson M, Brophy J (December 2001). "Community health profile of Windsor, Ontario, Canada: anatomy of a Great Lakes area of concern" (http://ehpnet1.niehs.nih.gov/members/2001/suppl-6/827-843gilbertson/gilbertson-full.html). *Environ. Health Perspect.*. 109 (Brogan &) **Suppl 6**: 827–43. doi:10.2307/3454645. JSTOR 3454645. PMC 1240618. PMID 11744501. .

[17] "Transit on Smog Days" (http://www.citywindsor.ca/001258.asp). Citywindsor.ca. . Retrieved 2012-01-02.

[18] "Parks and Facility Operations" (http://www.citywindsor.ca/000052.asp). City of Windsor. . Retrieved January 21, 2007.

[19] Selected Ethnic Origin for Windsor, 2001 (http://www12.statcan.ca/english/census01/products/highlight/ETO/Table1.cfm?Lang=E& T=501&GV=4&GID=3537039&Prov=35). Statistics Canada. Retrieved on 17 April 2009.

[20] Religion for Windsor, 2001 (http://www12.statcan.ca/english/census01/products/highlight/Religion/Page.cfm?Lang=E&Geo=CSD& View=3b&Table=1&Code=3537039&Sort=2&B1=Windsor &B2=Counts&B3=35). Statistics Canada. Retrieved on 17 April 2009.

[21] City of Windsor. *Demographics*. Available online at: http://www.citywindsor.ca/000503.asp

[22] *National Post*. "2001 census analysis: Highlights" Available online at: http://www.canada.com/nationalpost/story. html?id=3ee543f5-8c6b-4de0-acea-b4fe7305a42f

[23] "Demographics" (http://www.citywindsor.ca/002358.asp). Citywindsor.ca. . Retrieved 2012-01-02.

[24] "Ethnocultural Portrait of Canada - Data table" (http://www12.statcan.ca/english/census06/data/highlights/ethnic/pages/Page. cfm?Lang=E&Geo=CSD&Code=35&Table=1&Data=Count&StartRec=1&Sort=2&Display=Page&CSDFilter=5000). 2.statcan.ca. 2010-10-06. . Retrieved 2012-01-02.

[25] "Visible Minorities and Ethnicity in Ontario" (http://www.fin.gov.on.ca/en/economy/demographics/census/cenhi6.html). Fin.gov.on.ca. . Retrieved 2012-01-02.

[26] "Age & Sex" (http://www12.statcan.ca/english/census01/products/highlight/AgeSex/Page.cfm?Lang=E&Geo=CSD&Code=0& View=1&Table=4a&StartRec=701&Sort=2&B1=Median&B2=Both). 2.statcan.ca. . Retrieved 2012-01-02.

[27] Parliament of Canada (website) "History of Federal Ridings since 1867" (http://www2.parl.gc.ca/Sites/LOP/HFER/hfer. asp?Language=E&Search=R). . Retrieved 17 July 2007.

[28] "By-law to redivide the wards in the City of Windsor" (http://www.citywindsor.ca/DisplayAttach.asp?AttachID=14802). . Retrieved 2012-01-02.

[29] Sandwich First Baptist Church (http://www.historicplaces.ca/en/rep-reg/place-lieu.aspx?id=13374&pid=0). *Canadian Register of Historic Places.*

[30] François Bâby House (http://www.historicplaces.ca/en/rep-reg/place-lieu.aspx?id=12401&pid=0). *Canadian Register of Historic Places.*

[31] "'City of Windsor' Rose" (http://www.helpmefind.com/rose/pl.php?n=57961&tab=1). Helpmefind.com. . Retrieved 2012-01-02.

[32] "Red Bull Air Race" (http://www.redbullairrace.com/cs/Satellite/en_air/Official-Red-Bull-Air-Race-Homepage/001238611393596). Red Bull Air Race. . Retrieved 2012-01-02.

[33] Windsor locks in Red Bull air races for two years (http://www.windsorstar.com/entertainment/Bull+Race+returns+Windsor/2472987/ story.html)

[34] "CJAM 91.5 Windsor / Detroit Campus Community Radio" (http://web2.uwindsor.ca/cjam/index2.html). Web2.uwindsor.ca. . Retrieved 2012-01-02.

[35] "Canada South Science City" (http://www.cssciencecity.com/). Cssciencecity.com. . Retrieved 2012-01-02.

[36] "Doctor's Taking Patients" (http://www.ecms.org/), Essex County Medical Society. Retrieved 16 July 2007.

[37] "aircanada.com" (http://www.aircanada.ca/). Aircanada.ca. 2006-02-09. . Retrieved 2012-01-02.

[38] PORT WINDSOR - About the Port (http://www.portwindsor.com/portw.html)

[39] "Suit settled in death that led to overpass" (http://www.gregmonforton.com/news/overpass.pdf) (PDF). . Retrieved 2012-01-02.

[40] Changchun City, China website (http://2007.changchun.jl.cn/yingwenban/yingwen_detail.jsp?ID=120501000000000000,13). Retrieved 2 July 2009.

[41] Coventry Twin Cities (Windsor (http://www.coventry.gov.uk/ccm/content/chief-executives-directorate/corporate-policy/ international-team/windsor.en). Retrieved 2 July 2009.

[42] L'Association socioculturelle Granby et ses villes jumelées (http://www.ville.granby.qc.ca/villes-jumelees/index.html). Retrieved 2 July 2009.

[43] Gunsan City Worldwide Sisterhood Cities (http://web.archive.org/web/20071023210759/http://www.gunsan.go.kr/english/info/ info0601.jsp). Retrieved 2 July 2009.

[44] Lublin's Partner and Friend Cities (http://www.um.lublin.eu/en/index.php?t=200&id=40909). Retrieved 2 July 2009.

[45] Mannheims Partnerstädte - von Bydgoszcz bis Zhenjiang (http://www.mannheim.de/io2/browse/webseiten/tourismus/partnerstaedte/ index_de.xdoc). Retrieved 2 July 2009.

[46] City of Windsor, Our Twin Cities (Las Vueltas) (http://www.citywindsor.ca/001430.asp?city=lasvueltas). Retrieved 2 July 2009.

[47] Città gemellate (Windsor) (http://www.comune.udine.it/opencms/opencms/release/ComuneUdine/cittavicina/turismo/gemellate/ amiche/?style=1). Retrieved 2 July 2009.

[48] http://www.saintsathletics.ca

[49] "Welcome To Windsor Rugby (Windsor Rogues Rugby)" (http://www.windsorrugby.com/). Windsorrugby.com. 2011-11-25. . Retrieved 2012-01-02.

[50] http://www.windsornationals.ca/

[51] http://www.centralmma.com/index.html

External links

- Windsor-Essex Parkway (http://www.weparkway.ca/)
- City of Windsor (http://www.citywindsor.ca/)
- Community of Windsor, Ontario - Community Living Resources Windsor ON, Canada (http://www. canadawindsor.com/)
- CBC Windsor (http://windsor.cbc.ca/)
- Cycle Windsor, includes map of bike network, in PDF format (http://www.cyclewindsor.ca/)
- Community Portal (http://www.windsor-essex.info/)
- Arts Council Windsor & Region (http://acwr.net/)
- Article reflecting on the decline of the automotive industry in the area, by Jorn Madslien, BBC (http://news.bbc. co.uk/2/hi/business/7185787.stm)
- Woodford, Arthur M. (2001). *This is Detroit 1701–2001*. Wayne State University Press. ISBN 0-8143-2914-4.

Solid-state_physics

Solid-state physics is the study of rigid matter, or solids, through methods such as quantum mechanics, crystallography, electromagnetism, and metallurgy. It is the largest branch of condensed matter physics. Solid-state physics studies how the large-scale properties of solid materials result from their atomic-scale properties. Thus, solid-state physics forms the theoretical basis of materials science. It also has direct applications, for example in the technology of transistors and semiconductors.

Background

Solid materials are formed from densely-packed atoms, which interact intensely. These interactions produce the mechanical (e.g. hardness and elasticity), thermal, electrical, magnetic and optical properties of solids. Depending on the material involved and the conditions in which it was formed, the atoms may be arranged in a regular, geometric pattern (crystalline solids, which include metals and ordinary water ice) or irregularly (an amorphous solid such as common window glass).

The bulk of solid-state physics theory and research is focused on crystals. Primarily, this is because the periodicity of atoms in a crystal — its defining characteristic — facilitates mathematical modeling. Likewise, crystalline materials often have electrical, magnetic, optical, or mechanical properties that can be exploited for engineering purposes.

The forces between the atoms in a crystal can take a variety of forms. For example, in a crystal of sodium chloride (common salt), the crystal is made up of ionic sodium and chlorine, and held together with ionic bonds. In others, the atoms share electrons and form covalent bonds. In metals, electrons are shared amongst the whole crystal in metallic bonding. Finally, the noble gases do not undergo any of these types of bonding. In solid form, the noble gases are held together with van der Waals forces resulting from the polarisation of the electronic charge cloud on each atom. The differences between the types of solid result from the differences between their bonding.

Crystal structure and properties

Many properties of materials are affected by their crystal structure. This structure can be investigated using a range of crystallographic techniques, including X-ray crystallography, neutron diffraction and electron diffraction.

An example of a close-packed lattice

The sizes of the individual crystals in a crystalline solid material vary depending on the material involved and the conditions when it was formed. Most crystalline materials encountered in everyday life are polycrystalline, with the individual crystals being microscopic in scale, but macroscopic single crystals can be produced either naturally (e.g. diamonds) or artificially.

Real crystals feature defects or irregularities in the ideal arrangements, and it is these defects that critically determine many of the electrical and mechanical properties of real materials.

The crystal lattice can vibrate. These vibrations are found to be quantised, the quantised vibrational modes being known as phonons. Phonons play a major role in many of the physical properties of solids, such as the transmission of sound. In insulating solids, phonons are also the primary mechanism by which heat conduction takes place. Phonons are also necessary for understanding the lattice heat capacity of a solid, as in the Einstein model and the later Debye model.

Electronic properties

Properties of materials such as electrical conduction and heat capacity are investigated by solid state physics. An early model of electrical conduction was the Drude model, which applied kinetic theory to the electrons in a solid. By assuming that the material contains immobile positive ions and an "electron gas" of classical, non-interacting electrons, the Drude model was able to explain electrical and thermal conductivity and the Hall effect in metals, although it greatly overestimated the electronic heat capacity.

Arnold Sommerfeld combined the classical Drude model with quantum mechanics in the free electron model (or Drude-Sommerfeld model). Here, the electrons are modelled as a Fermi gas, a gas of particles which obey the quantum mechanical Fermi-Dirac statistics. The free electron model gave improved predictions for the heat capacity of metals, however, it was unable to explain the existence of insulators.

The nearly-free electron model is a modification of the free electron model which includes a weak periodic perturbation meant to model the interaction between the conduction electrons and the ions in a crystalline solid. By introducing the idea of electronic bands, the theory explains the existence of conductors, semiconductors and insulators.

The nearly-free electron model rewrites the Schrödinger equation for the case of a periodic potential. The solutions in this case are known as Bloch states. Since Bloch's theorem applies only to periodic potentials, and since unceasing random movements of atoms in a crystal disrupt periodicity, this use of Bloch's theorem is only an approximation, but it has proven to be a tremendously valuable approximation, without which most solid-state physics analysis would be intractable. Deviations from periodicity are treated by quantum mechanical perturbation theory.

Modern research in solid state physics

Current research topics in solid state physics include:

- Quasicrystals
- Spin glass

References

- Neil W. Ashcroft and N. David Mermin, *Solid State Physics* (Harcourt: Orlando, 1976).
- Charles Kittel, *Introduction to Solid State Physics* (Wiley: New York, 2004).
- H. M. Rosenberg, *The Solid State* (Oxford University Press: Oxford, 1995).
- *Out of the Crystal Maze. Chapters from the History of Solid State Physics*, ed. Lillian Hoddeson, Ernest Braun, Jürgen Teichmann, Spencer Weart (Oxford: Oxford University Press, 1992).
- M. A. Omar, *Elementary Solid State Physics* (Revised Printing, Addison-Wesley, 1993).

Moscow_Institute_of_Physics_and_Technology

<table>
<tr><td colspan="2" align="center">Moscow Institute of Physics and Technology (State University)</td></tr>
<tr><td colspan="2" align="center">Московский Физико-Технический институт (государственный университет)</td></tr>
<tr><td>Motto</td><td>Sapere aude (Dare to know; Dare to be wise)</td></tr>
<tr><td>Established</td><td>1946</td></tr>
<tr><td>Type</td><td>Public</td></tr>
<tr><td>Rector</td><td>Nikolay Kudryavtsev</td></tr>
<tr><td>Undergraduates</td><td>3319</td></tr>
<tr><td>Postgraduates</td><td>1758</td></tr>
<tr><td>Location</td><td>Dolgoprudny, Moscow, Zhukovsky, Russia</td></tr>
<tr><td>Website</td><td>www.mipt.ru [1], www.phystech.edu [2]</td></tr>
</table>

Moscow Institute of Physics and Technology (State University) (Russian: Московский Физико-Технический институт (государственный университет)), abbreviated **MIPT**, **MIPT (SU)** or informally **Phystech** (alternative transliterations: MFTI, Fizteh; МФТИ, Физтех) is a leading Russian university, originally established in the Soviet Union. It prepares specialists in theoretical and applied physics, applied mathematics, and related disciplines. It is sometimes referred to as "the Russian MIT."

MIPT is famous in the countries of the former Soviet Union, but is less known abroad. This is largely due to the specifics of the MIPT educational process (see "Phystech System" below). University rankings such as The Times Higher Education Supplement are based primarily on publications and citations. With its emphasis on practical research in the educational process, MIPT "outsources" education and research beyond the first two or three years to institutions of the Russian Academy of Sciences. MIPT's own faculty is relatively small, and many of its distinguished lecturers are visiting professors from those institutions. Student research is typically performed outside of MIPT, and research papers do not identify the authors as MIPT students. This effectively hides MIPT from the academic radar, an effect not unwelcome during the Cold War era when leading scientists and engineers of the Soviet arms and space programs studied there.

The word "phystech," without the capital P, is also used in Russian to refer to Phystech students and graduates.

The main MIPT campus is located in Dolgoprudny,[3] a northern suburb of Moscow. However the Aeromechanics Department is based in Zhukovsky, a suburb south-east of Moscow.

History

In late 1945 and early 1946, a group of prominent Soviet scientists, including in particular the future Nobel Prize winner Pyotr Kapitsa, lobbied the government for the creation of a higher educational institution radically different from the type established in the Soviet system of higher education. Applicants, carefully selected by challenging examinations and personal interviews, would be taught by, and work together with, prominent scientists. Each student would follow a personalized curriculum created to match his or her particular areas of interest and specialization. This system would later become known as the *Phystech System*.

The main building of MIPT

In a letter to Stalin in February 1946, Kapitsa argued for the need for such a school, which he tentatively called the *Moscow Institute of Physics and Technology*, to better maintain and develop the country's defense potential. The institute would follow the principles outlined above, and was supposed to be governed by a board of directors of the leading research institutes of the USSR Academy of Sciences. On March 10, 1946, the government issued a decree mandating the establishment of a "College of Physics and Technology" (Russian: Высшая физико-техническая школа).[4]

For unknown reasons, the initial plan came to a halt in the summer of 1946. The exact circumstances are not documented, but the common assumption is that Kapitsa's refusal to participate in the atomic bomb project, and his disfavor with the government and communist party that followed, cast a shadow over an independent school based largely on his ideas. Instead, a new government decree was issued on November 25, 1946 establishing the new school as a Department of Physics and Technology within Moscow State University. November 25 is celebrated as the date of MIPT's founding.[5]

MIPT campus

The four oldest residence halls are across the street from the academic buildings.

Kapitsa foresaw that within a traditional educational institution, the new school would encounter bureaucratic obstacles, but even though Kapitsa's original plan to create the new school as an independent organization did not come to fruition exactly as envisioned, its most important principles survived intact. The new Department enjoyed considerable autonomy within Moscow State University. Its facilities were in Dolgoprudny (the two buildings it occupied are still part of the present day campus), away from the MSU campus. It had its own independent admissions and education system, different from the one centrally mandated for all other universities. It was headed by the MSU "vice rector for special issues"--a position created specifically to shield the department from the University management.

As Kapitsa expected, the special status of the new school with its different "rules of engagement" caused much consternation and resistance within the university. The immediate cult status that Phystech gained among talented young people, drawn by the challenge and romanticism of working on the forefront of science and technology, and on projects of "government importance," many of them classified, made it an untouchable rival of every other school in the country, including MSU's own Department of Physics. At the same time, the increasing disfavor of Kapitsa with the government (in 1950 he was essentially under house arrest), and anti-semitic repressions of the late 1940s made Phystech an easy target of intrigues and accusations of "elitism" and "rootless cosmopolitanism." In the summer of 1951, the Phystech department at MSU was shut down.[6]

A group of academicians, backed by Air Force general Ivan Fedorovich Petrov, who was a Phystech supporter influential enough to secure Stalin's personal approval on the issue, succeeded in re-establishing Phystech as an independent institute. On September 17, 1951, a government decree re-established Phystech as the Moscow Institute of Physics and Technology.[7]

Apart from Kapitsa, other prominent scientists who taught at MIPT in the years that followed included Nobel prize winners Nikolay Semyonov, Lev Landau, Alexandr Prokhorov, Vitaly Ginzburg; and Academy of Sciences members Sergey Khristianovich, Mikhail Lavrentiev, Mstislav Keldysh, Sergey Korolyov, and Boris Rauschenbach. MIPT alumni include Andre Geim and Konstantin Novoselov, the 2010 winners of the Nobel Prize for Physics.[8]

The Phystech System

The following is a summary of the key principles of the Phystech System, as outlined by Kapitsa in his 1946 letter arguing for the founding of MIPT:

- Rigorous selection of gifted and creative young individuals.
- Involving leading scientists in student education, in close contact with them in their creative environment.
- An individualized approach to encourage the cultivation of students' creative drive, and to avoid overloading them with unnecessary subjects and rote learning common in other schools and necessitated by mass education.
- Conducting their education in an atmosphere of research and creative engineering, using the best existing laboratories in the country.

In its implementation, the Phystech System combines highly competitive admissions, extensive fundamental education in mathematics, as well as theoretical and experimental physics in the undergraduate years, and immersion in research work at leading research institutions of the Russian Academy of Sciences starting as early as the second or third year.

Departments

The institute has eleven departments, ten of them with an average of 80 students admitted annually into each.[9]

- Radio Engineering and Cybernetics [10]
- General and Applied Physics [11]
- Aerophysics and Space Research [12]
- Molecular and Biological Physics [13]
- Physical and Quantum Electronics [14]
- Aeromechanics and Flight Engineering
- Applied Mathematics and Management [15]
- Problems of Physics and Power Engineering [16]
- Innovation and High Technology [17]
- Nano-, Bio-, Information and Cognitive Technologies [18]

The eleventh faculty, Information Business Systems [19], offers only a Master's programme and accepts students with Bachelor's degrees from other faculties or other institutes.

Admissions

Most students apply to MIPT immediately after graduating from high school at the age of 17. Child prodigies are occasionally admitted at a younger age after skipping grades in school. Because admission is competitive, some of those who are not admitted reapply in subsequent years.

Traditionally, applicants were required to take written and oral exams in both mathematics and physics, write an essay, and have an interview with the faculty. The interview has always been an important part of the selection process. Sometimes an applicant with lower exam grades could be admitted, and one with higher grades rejected, based solely on the interview results.

In recent years, oral exams have been eliminated, but the interview remains an important part of the selection process.

The strongest performers in national physics and mathematics competitions and IMO/IPhO participants are granted admission without exams, subject only to the interview.

In accordance with the traditions of the Soviet education system, education at MIPT is free for most students. Further, students receive small scholarships (as of 2009, \$75–\$90 per month, depending on the student's performance), and effectively free housing on campus, which allows them to study full time.

Education

It normally takes six years for a student to graduate from MIPT. The curriculum of the first three years consists exclusively of required courses, with emphasis on mathematics, physics, and English. There are no significant curriculum differences between the departments in the first three years. A typical course load during the first and second years can be over 48 hours a week, not including homework. Classes are taught five days a week, beginning at 9:00 am or 10:30 am, and continuing until 5:00 pm, 6:30 pm, or 8:00 pm. Most subjects include a combination of lectures and seminars (problem-solving study sessions in smaller groups) or laboratory experiments. Lecture attendance is optional, while seminar and lab attendance affects grades.

A student studying the class schedule

MIPT follows a semester system. Each semester includes 15 weeks of instruction, two weeks of finals, and then three weeks of oral and written exams on the most important subjects covered in the preceding semester.

Starting with the third year, the curriculum matches each student's area of specialization, and also includes more elective courses. Most importantly, starting with the third year, students begin work at *base institutes* (or "base organizations," usually simply called *bases*). The bases are the core of the Phystech system. Most of them are research institutes, usually belonging to the Russian Academy of Sciences. At the time of enrollment, each student is assigned to a base that matches his or her interests. Starting with the third year, a student begins to commute to their base regularly, becoming essentially a part-time employee. During the last two years, a student spends 4–5 days a week at their base institute, and only one day at MIPT.

The base organization idea is somewhat similar to an internship in that students participate in "real work." However, the similarity ends there. All base organizations also have a curriculum for visiting students, and besides their work, the students are required to take those classes and pass exams. In other words, a base organization is an extension of MIPT, specializing in each particular student's area of interests.

While working at the base organization, a student prepares a thesis based on his or her research work and presents ("defends") it before the Qualification Committee consisting of both MIPT faculty and the base organization staff. Defending the thesis is a requirement for graduation.

Base organizations

As of 2005, MIPT had 103 base organizations. The following list of institutes is currently far from being complete:

- Institute for Information Transmission Problems RAS
- Institute for Nuclear Research RAS
- Institute for Physical Problems
- Institute for Problems in Mechanics RAS
- Institute for Spectroscopy Russian Academy of Sciences
- Institute for Theoretical and Experimental Physics
- Institute of Biochemical Physics RAS
- Engelhardt Institute of Molecular Biology RAS
- Shirshov Institute of Oceanology
- Institute of Molecular Genetics RAS
- Institute of Numerical Mathematics RAS
- Institute of Problems of Chemical Physics RAS
- Institute of Radio Engineering and Electronics of RAS
- Institute of Solid State Physics RAS
- Joint Institute for Nuclear Research
- Institute of Synthetic Polymer Materials RAS
- Institute for High Energy Physics
- Kurchatov Institute (formerly Kurchatov Institute of Atomic Energy)
- Lebedev Institute of Physics RAS (FIAN)
- Lebedev Institute of Precision Mechanics and Computer Engineering
- Landau Institute for Theoretical Physics
- N.N. Andreyev Acoustics Institute
- N.N. Semenov Institute of Chemical Physics, RAS
- Shubnikov Institute of Crystallography RAS
- Space Research Institute RAS
- Steklov Institute of Mathematics
- Gromov Flight Research Institute
- Zhukovsky Central Aerohydrodynamic Institute
- Nuclear Safety Institute of RAS (IBRAE)
- and a number of OKBs (experimental design bureaux)

In addition, a number of Russian and Western companies act as base organizations of MIPT. These include:

- 1C Company
- ABBYY
- Competentum Group or Physicon
- NPMP "Concept Consulting" [20]
- Intel
- IPG Photonics
- Kraftway
- MetaSythesis
- Paragon Software Group
- S.P. Korolev Rocket and Space Corporation Energia
- SWsoft
- Yandex

Degrees and reputation

Before 1998, students could graduate only after completing the full six-year curriculum and defending their thesis. Upon successful graduation, they were awarded a specialist degree in Applied Mathematics and Physics and, beginning in the early 1990s, a Master's degree in Physics.

Since 1998, students have been awarded a Bachelor's degree diploma after four years of study and the defense of a Bachelor's "qualification work" (effectively a smaller and less involved version of the Master's thesis). An estimated 90% of students continue their education after receiving this diploma to complete the full six-year curriculum and receive the Master's degree.

The complete course of education at MIPT takes six years to complete, just like an American Bachelor's degree followed by a Master's degree. However, MIPT graduates usually view their training as effectively higher than an American M.S. in Physics. The MIPT curriculum is, indeed, considerably more extensive compared to an average American college.[21] In addition, American M.S. programs usually focus more on classroom education and less on research. There is an opinion that an MIPT specialist/Master's diploma may be roughly equivalent to an American Ph.D. in physics[22] —possibly an undue generalization which, however, may be true in some cases.

Traditional university rankings are based on the universities' research output and prizes won by faculty.[23] In contrast, many distinguished professors teaching at MIPT are officially on staff at the base institutes (see above) rather than MIPT itself. Student research work is also typically carried out outside of MIPT, and published research results do not mention MIPT. In effect, many MIPT professors are not considered as such for the rankings, and student research is not earning any ranking points for MIPT.

Demographics

About 15% of all students are residents of Moscow and nearly the same are from Moscow region; the rest come from all over the former Soviet Union. Most out-of-town students live in the dormitory on campus for at least the first 4 years. Many senior students move to another dormitory in Moscow, while some either move to base institute dormitories or rent apartments.

The student population is almost exclusively male, with the female/male ratio in a department rarely exceeding 15% (seeing 2-3 women in a class of 80 is not uncommon). Although in recent years this situation has changed and in 2009 more than 20% of first year students were females.[24]

There are no reliable statistics on the careers of MIPT graduates. Prior to the collapse of Soviet Union, most MIPT graduates continued research at their base institutes or found jobs in OKBs. Nowadays, many graduates become business people or software engineers. Some, especially high-performing students of prestigious departments (e.g. DGAP, DCAM), go on to get post-graduate degrees from foreign universities. In the past, some students were known to have been admitted into Ph.D. programs of American universities as early as after their 3rd year of education. Many MIPT alumni hold faculty positions in the world's top Universities, including Harvard, MIT, Columbia, Stanford, Brown, and University of Chicago.

Famous faculty and alumni

- Alexander Abramov - founder of Evraz Group, #137 on the Forbes list
- Boris Aleshin - deputy prime minister in Russian government (2003–2004), president of AvtoVAZ (2007–2009), general director of TsAGI (2009-)
- Boris Babaian - a pioneer of Russian supercomputers, an Intel Fellow 2004[25] and software architect
- Yuri Baturin - former Russian head of national security, cosmonaut (1998 and 2001 missions)
- Oleg Belotserkovsky - rector of MIPT (1962–1987), prominent mathematician and mechanician
- Andrei Bolibrukh - a mathematician who solved Hilbert's twenty-first problem in 1989[26]
- Nikolai Borisovich Delone - a physicist who discovered multiphoton ionization.
- Alexander V Frolov - CEO of Evraz Group, #390 on the Forbes list
- Andrey Geim - discoverer of graphene, gecko tape, and levitating frogs; Fellow of the Royal Society, Nobel Prize in physics, 2010
- Konstantin Novoselov - Nobel Prize in physics for graphene research, 2010
- Vitaly Ginzburg - prominent physicist, Nobel Prize 2003,[27] co-developer of the Soviet H-bomb
- Yurij Ionov - discovered genome instability as a mechanism in colonic carcinogenesis[28]
- Aleksandr Kaleri - cosmonaut, spent 609 days on the Mir and ISS space stations
- Pyotr Kapitsa - discovered superfluidity,[29] Nobel Prize 1978[30]
- Leonid Khachiyan - famous for his Ellipsoid method for linear programming, Fulkerson Prize (1982)
- Mikhail Kirpichnikov - Russian Science & Technology Minister (1998–2000), dean of Biology at MSU (2006-)
- Alex Konanykhin - Entrepreneur, former banker, former Russian oligarch, with political asylum in USA.
- Nikolay Kudryavtsev - rector of MIPT (1997-), director of Schlumberger (2007-)
- Lev Landau - prominent Russian physicist, Nobel Prize 1962[31]
- Sergei Lebedev - invented MESM (1950) and BESM (1953) mainframe computers
- Alexander Migdal - defined 2D quantum gravity,[32] 2D/3D visualization software and internet entrepreneur
- Sergey Nikolsky - prominent Russian mathematician
- Alexander Polyakov - quantum field theory classics,[33] [34] [35] [36] [37] Dirac'86[38] and Lorentz'94 Medals
- Alexandr Prokhorov - a co-inventor of the laser, Nobel Prize 1964[39]
- Boris Rauschenbach - rocket scientist in control engineering, responsible for the first photographs of the far side of the Moon (1959)
- Boris Saltykov - Russian Minister of Science and Technology (1991–1996)
- Aleksandr Serebrov - cosmonaut, 373 days in outer space (four flights)
- Nikolay Semyonov - best known for his work on chain reactions, Nobel Prize 1956[40] in chemistry
- Natan Sharansky - Israeli Cabinet Minister (1996–2005), US Congressional Gold Medal (1986)
- Mikhail Shifman - non-perturbative QCD classics,[41] [42] Sakurai Prize (1999), Lilienfeld Prize (2006)
- Volodymyr Shkidchenko - Defense Minister of Ukraine (2003–2004), four-star general of the Army
- Rashid Sunyaev - an author of the Sunyaev-Zel'dovich effect and a model of black holes[43]
- Victor Veselago - put forward a theory[44] for metamaterials of the 21st century in 1967
- Alexander Zamolodchikov - quantum field theory classics[34] [36] [45]
- Dmitry Zelenin - governor of Tverskaya Oblast (2004–2011)

References

[1] http://www.mipt.ru/

[2] http://www.phystech.edu/

[3] MIPT on Google Map (http://maps.google.com/maps?t=k&hl=en&om=1&ll=55.929647,37.52287&spn=0.005319,0.007317)

[4] "Повесть древних времён или предыстория Физтеха", Ch 3 (http://potential.org.ru/bin/view/Home/ArtDt200503051018PH5J3) by N. V. Karlov.

[5] "Повесть древних времён или предыстория Физтеха", Ch 4 (http://potential.org.ru/bin/view/Home/ArtDt200504271517PH5J4) by N.V. Karlov.

[6] "Повесть древних времён или предыстория Физтеха", Ch 6 (http://potential.org.ru/bin/view/Home/ArtDt200506141012PH5J6) by N.V. Karlov.

[7] "Повесть древних времён или предыстория Физтеха", Ch 7 (http://potential.org.ru/bin/view/Home/ArtDt200507231448PH5J7) by N.V. Karlov.

[8] The Nobel Prize in Physics 2010 (http://nobelprize.org/nobel_prizes/physics/laureates/2010/)

[9] 2006 Admission Statistics (in Russian) (http://www.mipt.ru/abitur/stat/stat2006.html)

[10] http://frtk.ru

[11] http://www.dgap.mipt.ru

[12] http://www.faki.fizteh.ru

[13] http://bio.fizteh.ru

[14] http://ffke.fizteh.ru

[15] http://fupm.fizteh.ru

[16] http://fpfe.fizteh.ru

[17] http://fivt.fizteh.ru

[18] http://www.fnti.kiae.ru/www.htm

[19] http://fibs.fizteh.ru/

[20] http://www.acconcept.ru/department/department-about/brief.html

[21] Phystech's Educational Approach (http://www.mipt.ru/eng/questions/approach.html)

[22] Academicians, Hierarchy and Titles in Russian Science, MIPT Web Site (http://www.mipt.ru/eng/questions/hierarchy.html)

[23] Shanghai Jao Tong University ranking methodology (http://ed.sjtu.edu.cn/rank/2005/ARWU2005Methodology.htm)

[24] MIPT 2009 admittance statistics (http://www.mipt.ru/abitur/stat/stat2009.html)

[25] Intel Fellow, Boris A. Babayan (http://www.intel.com/pressroom/kits/bios/bbabayan.htm)

[26] Bolibrukh AA (1995). *21st Hilbert Problem for Linear Fuchsian Systems*. Amer Mathematical Society. ISBN 0821804669.

[27] The Nobel Prize in Physics 2003 (http://nobelprize.org/nobel_prizes/physics/laureates/2003/press.html)

[28] Ionov Y, Peinado MA, Malkhosyan S, Shibata D, Perucho M (1993). "Ubiquitous somatic mutations in simple repeated sequences reveal a new mechanism for colonic carcinogenesis". *Nature* **363** (6429): 558–61. Bibcode 1993Natur.363..558I. doi:10.1038/363558a0. PMID 8505985.

[29] Kapitza P (1938). "Viscosity of liquid helium below the λ-point" (http://www.nature.com/physics/looking-back/superfluid/index.html). *Nature* **141** (3558): 74. Bibcode 1938Natur.141...74K. doi:10.1038/141074a0. .

[30] The Nobel Prize in Physics 1978 (http://nobelprize.org/nobel_prizes/physics/laureates/1978/press.html)

[31] The Nobel Prize in Physics 1962 (http://nobelprize.org/nobel_prizes/physics/laureates/1962/press.html)

[32] Gross DJ, Migdal AA (1990). "Nonperturbative two-dimensional quantum gravity". *Phys. Rev. Lett.* **64** (2): 127–30. Bibcode 1990PhRvL..64..127G. doi:10.1103/PhysRevLett.64.127. PMID 10041657.

[33] Gubser SS, Klebanov IR, Polyakov AM (1998). "Gauge theory correlators from non-critical string theory". *Phys. Lett. B* **428** (1–2): 105–14. arXiv:hep-th/9802109. Bibcode 1998PhLB..428..105G. doi:10.1016/S0370-2693(98)00377-3.

[34] Belavin AA, Polyakov AM, Zamolodchikov AB (1984). "Infinite conformal symmetry in two-dimensional quantum field theory". *Nucl. Phys. B* **241** (2): 333–80. Bibcode 1984NuPhB.241..333B. doi:10.1016/0550-3213(84)90052-X.

[35] Polyakov AM (1981). "Quantum geometry of bosonic strings". *Phys. Lett. B* **103** (3): 207–10. Bibcode 1981PhLB..103..207P. doi:10.1016/0370-2693(81)90743-7.

[36] Knizhnik VG, Polyakov AM, Zamolodchikov AB (1988). "Fractal structure of 2d—quantum gravity". *Mod. Phys. Lett. A* **3** (8): 819–26. Bibcode 1988MPLA....3..819K. doi:10.1142/S0217732388000982.

[37] Polyakov AM (1977). "Quark confinement and topology of gauge theories". *Nucl. Phys. B* **120** (3): 429–58. Bibcode 1977NuPhB.120..429P. doi:10.1016/0550-3213(77)90086-4.

[38] Dirac Medallists 1986 (http://prizes.ictp.it/Dirac/DiracMedal86.html)

[39] The Nobel Prize in Physics 1964 (http://nobelprize.org/nobel_prizes/physics/laureates/1964/press.html)

[40] The Nobel Prize in Chemistry 1956 (http://nobelprize.org/nobel_prizes/chemistry/laureates/1956/press.html)

[41] Shifman MA, Vainshtein AI, Zakharov VI (1979). "QCD and resonance physics: The ρ-ω mixing". *Nucl. Phys. B* **147** (5): 519–34. Bibcode 1979NuPhB.147..519S. doi:10.1016/0550-3213(79)90024-5.

[42] Shifman MA, Vainshtein AI, Zakharov VI (1979). "QCD and resonance physics. Applications". *Nucl. Phys. B* **147** (5): 448–518. Bibcode 1979NuPhB.147..448S. doi:10.1016/0550-3213(79)90023-3.

[43] Shakura NI, Syunyaev RA; Sunyaev (1973). "Black holes in binary systems. Observational appearance". *Astron. Astrophys.* **24**: 337–55. Bibcode 1973A&A....24..337S.

[44] Veselago VG (1968). "The electrodynamics of substances with simultaneously negative values of ε and μ". *Sov. Phys. Usp.* **10** (4): 509–14. Bibcode 1968SvPhU..10..509V. doi:10.1070/PU1968v010n04ABEH003699.

[45] Knizhnik VG, Zamolodchikov AB (1984). "Current algebra and Wess-Zumino model in two dimensions". *Nucl. Phys. B* **247** (1): 83–103. Bibcode 1984NuPhB.247...83K. doi:10.1016/0550-3213(84)90374-2.

External links

- Official website (http://www.phystech.edu) (English)
- Official website (http://www.mipt.ru/) (Russian)

Article Sources and Contributors

Roman_Maev *Source*: http://en.wikipedia.org/w/index.php?title=Roman_Maev *Contributors*: 1ForTheMoney, Alex Bakharev, Baumfreund-FFM, Bearcat, Ben Ben, CommonsDelinker, Headbomb, Mmancini81, Nyttend, Shawn in Montreal, Sk741, 1 anonymous edits

Nikolay_Basov *Source*: http://en.wikipedia.org/w/index.php?title=Nikolay_Basov *Contributors*: Alex Bakharev, Altenmann, Amanda.nelson12, Amikake3, Andrei Stroe, Andy.goryachev, Ashot Gabrielyan, Azov, Beetstra, Book-monitor, Ceancata, Chicheley, Cmapm, Corrigendas, D6, Deerstop, Dicklyon, Dirac1, Emerson7, Folks at 137, Gcm, Greyhood, Gulmammad, Headbomb, HennessyC, J.smith, James Gunasekera, JdH, Jimmyeatskids, Joao Xavier, John, KNewman, Ksnow, LarRan, Liilliil, Lupo, MZMcBride, Margoz, Mhym, Mnmngb, Monegasque, Netsnipe, Nguoimay, Omnipaedista, PasswordUsername, Pedron, Phe, Rich Farmbrough, Rjwilmsi, Rocastelo, Rsabbatini, Sannita, ScottDavis, Sk741, Srbauer, Srleffler, Tamtamar, Todowd, USA 192131119, Ulrika F., Unyoyega, Wuhwuzdat, Zloyvolsheb, 20 anonymous edits

Nobel_Prize_in_Physics *Source*: http://en.wikipedia.org/w/index.php?title=Nobel_Prize_in_Physics *Contributors*: 62.253.64.xxx, A bit iffy, A. di M., A2Kafir, AMK152, Aatomic1, Adam78, Agathoclea, Ahoerstemeier, Alansohn, Alexnye, Amoruso, Ancheta Wis, Andre Engels, Andrius.v, Andrwsc, Anonymous Dissident, Aporwitz, Aranea Mortem, Art Carlson, Attilios, Aulis Eskola, AxelBoldt, Axelv, BRG, Ben davison, Benhocking, Bevo, Big Brother 1984, BigMo1, Billthefish, Bobblewik, Bobryuu, Brandmeister, Brion VIBBER, Butseriouslyfolks, C.Fred, Cantus, Canuckian89, Cap.fwiffo, Charles Matthews, Chowbok, Chrishmt0423, Christian List, ChristopherWillis, CieloEstrellado, Clarityfiend, Cmapm, Cntras, Cometstyles, Conversion script, Coyets, Curps, Cyanidethistles, Cyp, D.H, Danny, David Kernow, Deb, Derek Ross, Dirac1933, Doug Bell, DrMikeF, Drgarden, DubaiTerminator, Dysepsion, EamonnPKeane, Eclecticology, Edgarisaballer, El C, Eliz81, Eman, Emerson7, Emilindito, Esuzu, Everyking, Evil Monkey, F1list, Fastfission, Fibonacci, Fram, Fredrik, Friendly Neighbour, Fryede, Ftunt hrei, Funandtrvl, Gabbe, Galoubet, Gareth Wyn, Garion96, Gene Nygaard, Giftlite, Gobonobo, Graham, Graham87, GregorB, Hapsala, Harp, Headbomb, Hemlock Martinis, Hfastedge, I am neuron, Il MusLiM HyBRiD II, Iantresman, Icairns, Itinerant1, Iwikonibla, Izbitzer, J.delanoy, JYOuyang, Jac16888, Jarszick, Jauhienij, Jay, Jayarathina, Jbgjbg, Jheald, Jiang, Joao Xavier, Jonathunder, Joyous!, Jredmond, KTC, Karaboom, Kinglouie324, Kopelitsa, Kpjas, Kransky, Ksnow, Kumkee, Kupirijo, Kutchkutch, Labongo, Landroo, Lcarscad, Lesadistic, Leszek Jańczuk, Lightmouse, Lihaas, Linas, Lockview22, Lookoo, Looxix, Lord Emsworth, Lowellian, MARKELLOS, MCMLXXXIII, MadAdam211, Manning Bartlett, Marek69, Masterpiece2000, Mav, Mawai, Mdd, Mekong Bluesman, Member, Mhwu, Mic, Michaelschmatz, Micoolio101, Modest Genius, Moose-32, Mormegil, Mxn, Myasuda, NYScholar, Nameneko, Nanobug, Nascigl, Ndkartik, Newone, Nigholith, Nisselua, Nitya Dharma, Nova Cygni, Npeters22, Nwt, Obradovic Goran, Olivier, Ornithologician, Oroso, Oskar Breach, OwenX, Panda, Patchouli, Pavaniprasanna, Phgao, Pit, Prashanthns, Pred, Prince Max (scientist), Psiacolyte, Qmwne235, Qwerty Binary, Qwertyus, RJHall, RachelBrown, RandomP, RasmusXF, Rbb 1181, RedWolf, RexNL, Reywas92, Rglovejoy, Rich Farmbrough, Rillian, Rjwilmsi, Rmhermen, Rnt20, Robertgreer, Saadkalaniya, Sak31122, Salsa Shark, Sceptre, Scorpion0422, SeNeKa, Search4Lancer, Shadowjams, Shanes, Sharkface217, Shizhao, Shlishke, Shreevatsa, Sliggy, Solitude, SpeedyGonsales, Spookee, Srleffler, Stevenscollege, Stone, TRBlom, TakuyaMurata, Tarawneh, Tarquin, Tasfhkl, Template namespace initialisation script, Templatehater, Tha prez, Thatha, The Anome, The undertow, The wub, Tim Starling, Timwi, Tjs2012, Tomas e, Uceboyx, Uncle Milty, UnitedStatesian, Uppland, Vald, Valentinian, Volker89, Wernher, WikiBahal, Wikipedian06, Wshun, XJamRastafire, Youandme, Youssefsan, Zeno Gantner, Шиманський Василь, , 379 anonymous edits

Alexander_Prokhorov *Source*: http://en.wikipedia.org/w/index.php?title=Alexander_Prokhorov *Contributors*: Adam Zábranský, Alex Bakharev, Aris Katsaris, Ashot Gabrielyan, Assedo, Azov, Book-monitor, Calmer Waters, Ceancata, ChicXulub, Chicheley, Chris93, Clivepollard, Closedmouth, Cmapm, Colonies Chris, Crazyf29, Cyp, D6, Dangshei, Dicklyon, Emerson7, Everyking, Ezhiki, Favonian, Folks at 137, Gcm, Georgecmu, Greyhood, Gvf, Hans castorp81, Headbomb, HennessyC, Hmains, J JMesserly, JackofOz, Jackyd101, Jcaragonv, Jimmyeatskids, Joao Xavier, KNewman, Kelson, Ksnow, Ktr101, LarRan, Margoz, Masterpiece2000, Materialscientist, Maximus Rex, Mic, Mnmngb, Netsnipe, Nevilley, Newtowiki2009, No Free Nickname Left, Phe, Piledhigheranddeeper, Plindenbaum, RS1900, Rjwilmsi, Romanm, Russavia, Salsa Shark, Silvonen, Sk741, Sluj, Srleffler, Stern, Tec15, XJamRastafire, Zloyvolsheb, راهب نی‌شن‌مه, 24 anonymous edits

Quantum_optics *Source*: http://en.wikipedia.org/w/index.php?title=Quantum_optics *Contributors*: A. B., Agge1000, Alansohn, AmarChandra, Asfarer, Assedo, Austin Maxwell, Bakrivan, Benevolentkoi, Beowulf333, Brazmyth, Charles Matthews, ChicXulub, Chub, Crmrmurphy, DARTH SIDIOUS 2, David Newton, DavidGrayson, Dekona, Dicklyon, Dirac1933, Dolbin, Eng.ahmedsamy, F=q(E+v^B), Fabrictramp, Gabroni, Gaius Cornelius, Gerd Breitenbach, GermanX, Gfutia, Grim23, HappyCamper, Harold f, Hess88, J S Lundeen, Jo3sampl, John Belushi, Knotwork, Laurascudder, LeonWhite, Linas, Lysdexia, Materialscientist, Matt McIrvin, Mechanical digger, Michael P. Barnett, MykReeve, Navasj, Nerone, Pgabolde, Pjvpjv, Pmokeefe, Pvvni, REwhite, RPaschotta, RobinK, Sanders muc, Sapphic, SchillerStephan, Sijokjoseph, Srleffler, Stanford96, Staz69uk, SteinAlive, Sylgeist, Synergy, Thingg, Thunderhead, Tomatoman, Waxigloo, WikiDan61, Zarniwoot, ZorkFox, 57 anonymous edits

Physicist *Source*: http://en.wikipedia.org/w/index.php?title=Physicist *Contributors*: 1836311903, 786knowledge, Aditya, Aksi great, Alan Liefting, Alansohn, Aleenf1, Alexander.stohr, Altenmann, Amcbride, Arthena, AzaToth, Bart133, Bbullot, Bduke, Bobianite, Bobo192, Brian0918, CalumH93, Chaiken, Cohesion, Columbia13, CommonsDelinker, Conversion script, Cybercobra, Death66613, December21st2012Freak, Dennis Estenson II, Denny, DerHexer, Doulos Christos, Drphysics, E235, Edivorce, Edwinstearns, El Cid, Elminster Aumar, EncycloPetey, Enormousdude, Epbr123, Ergative rlt, Escape Orbit, Etniesskater10, Ewlyahoocom, Fasach Nua, Fastfission, Flewis, G. Moore, Gala.martin, Galoubet, Gganges, Giftlite, Gjp23, Globalsolidarity, Goethean, Harriv, Headbomb, Helix84, Husond, IncognitoErgoSum, Ingjald Pilskog, Ivan Bajlo, JBKramer, JaGa, Jagged 85, Jaraalbe, Jauhienij, Jeffrey Mall, Jholman, Jiang, Josephcunningham, JoshHoward77, Just plain Bill, Justin W Smith, Karol Langner, Katalaveno, King of Hearts, Kkm010, Kostisl, Krille, KuduIO, Kungfuadam, LarRan, LaughingVulcan, Laurascudder, Lauren.e.puretz, Lightmouse, MER-C, Marek69, Mark91, Markaci, MassimoAr, Maurice Carbonaro, Maurreen, Mav, Melburnian, Mephistophelian, MichaelMaggs, Monfornot, Motor, Muhsinim, Mxn, Natalie Erin, NuclearWarfare, Ollie214345332312, Oxymoron83, Peregrine981, Peterlin, Phgao, PhilKnight, Philip Trueman, Physis, Piet Delport, Pigsonthewing, Polyamorph, Poor Yorick, RG2, Rifeldeas, Rjwilmsi, Robby, Robminchin, Rsm99833, Ryoutou, Samuel Blanning, SciCorrector, Scottfisher, Shiaoster, Shoeofdeath, Sibian, Sophus Bie, Stevenreed037, Stormwriter, Suaheli, Tachyon502, Techraj, The Anome, The Thing That Should Not Be, Thingg, Tide rolls, Tiefighter, Tnxman307, Tobby72, Tr-the-maniac, Tuttiverdi, Ukexpat, Uvainio, VanishedUser314159, Voxii, WadeSimMiser, Wesselbindt, Wiki alf, Willworkforicecream, Winston365, WojPob, Wolfkeeper, XJamRastafire, Yamamoto Ichiro, Zowie, احمدغامدي.24, ساىئ, 288 anonymous edits

Charles_Hard_Townes *Source*: http://en.wikipedia.org/w/index.php?title=Charles_Hard_Townes *Contributors*: Alan Liefting, Amillar, Another Philosopher, ArglebargleIV, Ashot Gabrielyan, Astuishin, BD2412, Badbilltucker, Benjaminevans82, Betacommand, Bolikesboys, Bunzil, Canadian Paul, CanisRufus, Captain Proton, Charles Matthews, ChicXulub, Chicheley, Cinik, Clivepollard, CommonsDelinker, Ctrl build, Cxbrx, D6, Davshul, Deb, Dewey Finn, Dhaluza, Dicklyon, Doctorcherokee, Drdavidhill, Dudleyt, Dumelow, Dysprosia, Ekabhishek, Elb2000, Emerson7, Epbr123, Equilibrial, Etacar11, Everyking, Firefly322, Fitzaubrey, Frank, Fredrik, Gaius Cornelius, GangofOne, Garion96, GcSwRhlc, Gcm, Gene Nygaard, Gershwinrb, Giftlite, Good Olfactory, Guychua, Gvf, Headbomb, HennessyC, Hephaestos, IRP, InverseHypercube, Ixfd64, JLaTondre, Jack Cox, Japanese Searobin, Jiuguang Wang, Jkaharper, Jrcla2, KSchutte, Karaboom, Katharineamy, Ketiltrout, Kkm010, Ksnow, Kumioko, LaszloWalrus, Lemeza Kosugi, Ljg377, Lockley, MWaller, Madcoverboy, Masterpiece2000, Mathiasrex, Mattbr, Maximus Rex, Meco, Mehulmalik, MessinaRagazza, Micru, Mnmngb, MoiraMoira, MuJami, NBeale, NJA, Newtowiki2009, Newyears, Oaktree b, PDH, PearlSt82, Pediainsight, Peter Ellis, Phe, Photokid2010, RS1900, RenoVSCid, Reywas92, Rglovejoy, Rich Farmbrough, Richard Arthur Norton (1958-), Rnt20, Roger Hui, Rror, Rsabbatini, Ryanluck, Rye1967, Shrubbery, Shtaif, Silvonen, Singlephoton, Skepper43, Snowolf, Socersam627, Spencerweart, Spoon!, Srbauer, Srleffler, SuperGirl, T. Anthony, Takwish, Tinlash, Twthmoses, Usctommytrojan, Valentinian, Vojvodaen, Vsmith, WOSlinker, Walker123, Wechselstrom, WingkeeLEE, Wmahan, WriterListener, راهب نی‌شن‌مه, 98 anonymous edits

Maser *Source*: http://en.wikipedia.org/w/index.php?title=Maser *Contributors*: 216.237.32.xxx, AGToth, Aaron Walkhouse, Accurizer, Adrigon, Alain Michaud, Aldie, Alison, AmuroNT1, Anchpop, Anthony Appleyard, Ashley Pomeroy, AstroNomer, Attilios, Bardic Nerd, Bci2, Bongwarrior, BrownHairedGirl, Charleca, Cholmes75, Cmapm, Commander, Conversion script, Crazilla, Creelbm, Cyberherbalist, Cypa, DaedalusZM, Deathregis, Deglr6328, DrBob, Drxenocide, Dzsi, Eloquence, Eumolpo, Eyreland, Fig wright, FigmentJedi, Fjörgynn, Franco3450, Funandtrvl, Gene Nygaard, George100, Gimboid13, Ginsengbomb, Glenn, Hans castorp81, Heron, Hmoraga, JHunterJ, Jeremyjackson, Jeremyvaunbaumseexthethird, Jim whitson, Jiminy pop, Jonjonmehigan, Joyous!, Jrockley, JustinWick, Khoikhoi, Kit Foxtrot, Kjkolb, Kkmurray, Kku, Kxdan13, Le Blue Dude, Lyle Swann, Mangina6969, Marudubshinki, Materialscientist, Mekong Bluesman, Mhazard9, Michael Devore, Micru, Mion, Monedula, Mythsearcher, Nightscream, Obsessive writer, Old Moonraker, Omegatron, Peter Vince, SchfiftyThree, Scubbo, Sertrel, Shaddack, Shaun F, Shawn Worthington Laser Plasma, Siafu, Snowdog, Spencerweart, Spidermax10, Srleffler, Stratocracy, Suruena, Syndicate, Tim!, Tomchiukc, Trojancowboy, Vanished user 47736712, Vgy7ujm, Vicki Rosenzweig, Viggle69, W2raphael, Wgfcrafty, Who, Wiccan Quagga, Yath, Zephalis, Zundark, 181 anonymous edits

Windsor,_Ontario *Source*: http://en.wikipedia.org/w/index.php?title=Windsor%2C_Ontario *Contributors*: -- April, 203.37.81.xxx, A-MAN, A3RO, Absalom89, Academic Challenger, Ace of Spades, Adam Bishop, Adolch, Advicemaster21, Aegis Maelstrom, AeomMai, Agha Nader, Ahoerstemeier, Alansohn, AlexRampaul, Alpha 4615, Amardeshbd, Amicuspublilius, Anas Salloum, Andy Marchbanks, Angela2109, Angr, Ante Aikio, AnthonyConway, Antiuser, Antonrojo, Aou, Arch26, ArchipotleSauce, Argyll Lassie, Arthree, Artspeak, Ary29, Atubeileh, Avenged Eightfold, Azuramanga1, B.Vandy, BCKILLa, BD2412, Bacl-presby, Baldwin.jim, Barkjon, Barticus88, Bcorr, Bearcat, Beetstra, Benjibol, Berton, Bidabadi, Big Bird, Big iron, Bigjohndallas, BilCat, Blackjays, Blackjays1, Bllasae, Blotto adrift, Bluedustmite, Bobblehead, Bobblewik, Bobby H. Heffley, Bobo192, Boneillhawk, Borgx, Brad Lombardo, Bradeos Graphon, BrendelSignature, Brenont, Brian Crawford, Brossow, Brusegadi, Bryabail, ByteofKnowledge, CALR, CMC, CRACK MONKEY13, CRKingston, Calabe1992, Can't sleep, clown will eat me, Canada3000, Canadian Bobby, CanisRufus, Capacch, Captinvideo, Catgut, Chamal N, Chanheigeorge, Chnou, Ciotog, Circeus, Classic Katarn, Cmr08, Coachforeman, Colonies Chris, Cometstyles, CommonsDelinker, Communitarian35, Conversion script, CookieWorld04, Crappyusername, CrazyC83, Cricket02, Cris10103, Criticalthinker, Cts240, D6, DMighton, Dana boomer, Danthemankhan, Darius Dhlomo, Dark Tichondrias, Data beagle a1a, Davert, David Kernow, Davidlak, DeadEyeArrow, Decimal10, Deconstructhis, Decumanus, Dekimasu, DerHexer, Derkasssss, Dgebel, Dgies, Dgrant, Dhartung, Diligent Terrier, Dkolodziej, Dl2000, Doganman, Dogru144, Dominguez, Doze, Dpm64, DrVenkman, DragonofFire, Dreamyshade, Drilnoth, Dschwen, Dully 16, Dunro, Dyfsunctional, Earl Andrew, Eclecticology, Ed g2s, Eelamstylez77, Eja2k, El C, Eliz81, Emarsee, Embee473, Enduser1950, Escapeartistsneverdie13, Eugene Girard, Everyking, Fages, Fancypants09, Farmerman, Farside268, Fastestdogever, FatM1ke, Faustian, Feydey, Flibirigit, Floydian, Flux88, Foobaz, Foxtrotman, Frankie816, Frankrig, Fratrep, Freakofnurture, Future Perfect at Sunrise, Fuzheado, Gaff, Gaius Cornelius, Galati, GallanoHJS, Garywill, Geo android, Gimboid13, Gokujoseitokai, Gracenotes, GraemeL, Green Giant,

Greenshed, Gregbaker, Grouf, Ground Zero, Groundling, Gunter Viezenz, HFHGH, HaJor, Haljackey, Hammer1289, Handicapper, HangingCurve, Hayden120, Hbayat, HelloAnnyong, HennessyC, Heyheyheyballday, Hick31, Hikaru79, Hmains, HomeFlower, Horbal, Horst-schlaemma, Howcheng, Htaccess, Hunting dog, Hydnjo, INkubusse, Ibagli, Ilws, Indefatigable, Infinitis24, Ingridi2, Intelligentsium, Iridescent, Irregulargalaxies, Ixfd64, J'onn J'onzz, J.delanoy, JPG-GR, JSDA, Jab843, Jack Cox, Jackrosenberg, JakeM700, Jc8025, Jd.101, JeLuF, Jeff schiller, Jemiller226, Jhendin, Jimmy socks, JimmyTrump79, Jluuwindsorca, JoanneB, Jogers, John, Jon Awbrey, Jonathan.s.kt, Jonathanbc, Joseph Solis in Australia, Joyous!, JuWiki2, Jusjih, JustAGal, KGasso, Kaitauth, Kalsermar, Kelapstick, Khazar, Kleinerkevin, Kootenayvolcano, Kory65, Kubigula, Kusunose, Kyle1278, L Kensington, LOL, Last apoc, Leafsman 67, Lexicon, Light Grenade, Lightmouse, Ling.Nut, Localsportswindsor, Loodog, Lwalt, MJCdetroit, Macker626, Macosx, Manutd7, Marara19, Marbod Egerius, Marek69, Mark K. Jensen, Marknen, Martarius, Martin451, Mastermtg, Mathpianist93, Matt 314, Mattsrevenge, Maury Markowitz, Mayumashu, Mboverload, McPhail, Mdb1370, Merkin77, Mertozoro, Mhking, MiLo28, Michiganinplay, Mikerussell, MilborneOne, Mindmatrix, Mindme, Mintguy, MissMassaro-xO, Mousky67, Mr Accountable, MrFish, Mrbrownn, Mrpick, Mswoc, MuZemike, Mwachna, Myanw, Mysid, NE2, NPrice, Nbaballer55555, NeoChaosX, NewEnglandYankee, Nhl4hamilton, Nika 243, Nmj, Noirestaurant, NormanEinstein, NorthernThunder, Notorious4life, Nyttend, Oaronuviss, Objectivesea, Ohnoitsjamie, Oo7565, Oreo Priest, Otav347, Ottre, Oxymoron83, P199, PKT, Padjet1, Pagingmrherman, Panfrie, Parkwells, Patrick, PaulHanson, PeRshGo, Pedx, Peruvianllama, Peter Grey, Peter243, Peterzahran, Phantomsteve, PhilKnight, Philippe, PhilthyBear, Pikks, Pinac, Pk bobb 1, Po' buster, Ponyo, Poor Yorick, Prasanaik, Press olive, win oil, Prince Diamond, Prodego, Qasrani, Qazxsw1234, Quadalpha, Quantumobserver, Qutezuce, R'n'B, RDBrown, RFBailey, RFD, RVJ, Raccoon Fox, RaccoonFox, Radagast, Ramsin187, Razorflame, Reach Out to the Truth, Recurring dreams, Rich Farmbrough, Rick Block, Rickyfusob, Rigind, RingtailedFox, Rjwilmsi, Rlquall, Rmhermen, Rob Russell, Robert talan, Rodii, Rosiestep, Roy86, Rubberbandman90, Rune.welsh, Rupertslander, Ryan392, Ryanamy83, S in mtl, ST47, Sammo, Samy23, Sango123, Saran81kid91, Sfan00 IMG, Shawn in Montreal, Sheba7, Shiviscs, Shizane, Shizukujapaneserice, Sietse Snel, Silpol, SimonP, Sjorford, Skater100, Skeezix1000, Some jerk on the Internet, Somercet, Spike Wilbury, Spliffy, Spmarshall42, Spylab, Srnec, Steam5, Stephenb, Stepheng3, Stevup, Stewacide, Stifle, Storkk, Strand, SunCreator, Tanweer Morshed, TayllLapo, Tbirdskelding, Teenwindsor08, Tellyaddict, The Lake Effect, The wub, Theatre Manager, Thecurran, Theed, Themightyquill, Thesot, Thirty-seven, Thiseye, Thisismydamnusername, Thomas Paine1776, Tiger888, Tim1965, Timc, Timothygetsschooled, Titorejre, Tkgd2007, Tlaratroi, Tlogmer, Tml32, Tpbradbury, Travelbird, Truthanado, Turian, Ucanlookitup, Ultra megatron, UnQuébécois, Uncle G, Uncle Milty, UndeadDawning, Unschool, Upfrontmagazine, Vegaswikian, Vgy7ujm, ViperSnake151, WODUP, Wangry, Warrior1867, Wavelength, Weirdstapler, WhisperToMe, WikHead, Wikiedit90210, Wikipelli, Wikiuser11malibu, Windsor's Communiy Museum, Windsorconnected, Windsordude, Windsorgurl, Windsorites, WolfmanSF, Ww2censor, Xous, Yupislyr, Zosopage, Zutroy Tibor, Zzyzx11, 1186 anonymous edits

Solid-state_physics *Source*: http://en.wikipedia.org/w/index.php?title=Solid-state_physics *Contributors*: ANDROBETA, Agemoi, ArnoLagrange, Austin Maxwell, Bgramkow, CYD, Chaiken, Clipjoint, Cool3, Costyn, Csmallw, Css, Cutler, CyborgTosser, Dchristle, Djr32, EagleFan, Edkarpov, El C, Eras-mus, FlorianMarquardt, Gcm, GermanX, GraemeL, Haljolad, Headbomb, Hongooi, Ibrahim-saeed, Ilyaroz, Interiot, Juhis, Karol Langner, Labizun, Linas, Magister Mathematicae, Marcok, Miranda, Mosaffa, MrOllie, Nk, Notinasnaid, Ozuma, Piotrek45, RA0808, Radagast83, Rmalloy, RockMagnetist, Romaioi, Rorro, Sbarnard, Srleffler, Stanford96, Steve Quinn, StewartMH, Tantalate, Template namespace initialisation script, Tim Starling, Tyrol5, Unmerklich, Wacko Guy, Wbm1058, Wimt, WingkeeLEE, Wthered, Yvwv, Zaheen, Zarniwoot, 57 anonymous edits

Moscow_Institute_of_Physics_and_Technology *Source*: http://en.wikipedia.org/w/index.php?title=Moscow_Institute_of_Physics_and_Technology *Contributors*: .:Ajvol:., 123Michael, Altenmann, Avn isan, Chaojoker, Cmapm, D6, Darena mipt, Docu, DrStrangeLove, Eclipsed, Edgar181, EncephalonSeven, Eplaton, Errandir, Everyking, Ezhiki, Freeroler, Gene s, Good Olfactory, Goudzovski, Gseryakov, HandsomeFella, Headbomb, Hmains, Humus sapiens, Ipsign, Itinerant1, JackofDiamonds1, Jeandré du Toit, Koavf, Latash, Martarius, Mhym, Mishas42, Mitvorot, Orderinchaos 2, Paul Furret, R'n'B, Rassul, Reflex Reaction, Rich Farmbrough, Rjwilmsi, Rossami, RoundPanda, Rsynnott, Salih, ScalarField, Sergeom, Sk741, Skybon, Steve carlson, Tassedethe, Tec15, Tikiwont, Varepsilon, Vlsergey, Vmenkov, Woohookitty, 146 anonymous edits

Image Sources, Licenses and Contributors